Geophysical Monograph Series

Including

IUGG Volumes
Maurice Ewing Volumes
Mineral Physics Volumes

GEOPHYSICAL MONOGRAPH SERIES

Geophysical Monograph 57
IUGG Volume 8

Evolution of Mid Ocean Ridges

John M. Sinton, Editor

American Geophysical Union
International Union of Geodesy and Geophysics

Geophysical Monograph/IUGG Series

Library of Congress Cataloging-in-Publication Data

Evolution of mid-ocean ridges.

   (Geophysical monograph ; 57) (IUGG ; v. 8)
   Based on IUGG Union Symposium 9 held during the 1987
IUGG General Assembly at Vancouver, Canada, and jointly
sponsored by IAVCEI and others.
   1. Sea—floor spreading—Congresses.   2. Mid-ocean
ridges—Congresses.   I. Sinton, John M.   II. IUGG Union
Symposium (9 : 1987 : Vancouver. B.C.)   III. Internation-
al Association of Vocanology and Chemistry of the Earth's
Interior.   IV. Series.   V. Series: IUGG (Series) ; v. 8.
QE511.7.E96   1989   551.46'08   89-18531
ISBN 0-87590-458-0

CONTENTS

This volume is an outgrowth of IUGG Union Symposium 9 held during the 1987 IUGG General Assembly at Vancouver, Canada. This symposium, jointly sponsored by IAVCEI, IASPEI, ICL and IAGA, consisted of 31 presentations ranging in subject matter from melt segregation and melt focusing processes beneath mid-ocean ridges, to the structures of oceanic crust and ophiolite analogues, morphological variations in the accretion process, the structural evolution of specific spreading ridge systems, the interplay between magmatism and rifting, and the chemical and thermal balances involved in mid-ocean ridge hydrothermal systems. Six of those papers have been expanded in the present volume.

These papers constitute several important advances in our understanding of the evolution of mid-ocean ridge systems. The recognition that transverse seismic anisotropy is an important characteristic of oceanic layer 2 (Fryer et al.) has profound implications for interpretations of crustal thicknesses based on seismic data, and appears to explain a longstanding enigma of marine seismology: the apparent thinning of upper crustal layers with age. An analysis of magnetic anomaly data and transform fault azimuths across the boundaries of the Pacific, Easter and Nazca plates (Naar and Hey) has resulted in the calculation of new, instantaneous plate motion models for a significant portion of the south Pacific plate boundaries, in addition to providing important constraints on the recent evolution of the Easter Microplate. A new kinematic model for the evolution of the Gorda Rise (Stoddard) reproduces the complex magnetic lineations of that area, and includes models for the genera-tion of the President Jackson seamount chain. Phase equilibria are used to constrain the nature of magmas parental to differentiated lavas of Icelandic rift zones (Thy); these magmas contrast significantly with those for several other spreading ridges, with implications for the melting regimes operating there. The final two papers are devoted to evaluations of the accretion process over relatively short time intervals. The use of bottom observations at Axial Seamount on the Juan de Fuca Ridge has allowed Zonenshain et al. to decipher the volcanic, tectonic and hydrothermal history of this area over the last 60,000 years. An even finer scale view of the accretion process is provided by Jacoby et al., in their assessment of the implications of geophysical and geodetic data for magma movement in the Krafla Rift Zone, Iceland since 1975.

Collectively, these papers should lead to better understanding of the process of accretion at mid-oceanic ridges, and the structure and evolution of the oceanic crust produced in the plate boundary zone. The efforts of the following reviewers are gratefully acknowledged for their contributions to improving the manuscripts in this volume:

R. L. Chase	M. Fisk	K. C. Macdonald	D. Walker
N. Christiansen	R. N. Hey	B. Minster	G. P. L. Walker
P. Einarsson	J. Karsten	G. M. Purdy	D. S. Wilson

John M. Sinton
Honolulu, Hawaii

Evolution of Mid Ocean Ridges

SEISMIC ANISOTROPY AND AGE-DEPENDENT STRUCTURE OF THE UPPER OCEANIC CRUST

Gerard J. Fryer, Daniel J. Miller[1], and Patricia A. Berge

Hawaii Institute of Geophysics, University of Hawaii at Manoa, Honolulu, Hawaii 96822

Abstract. Seismic anisotropy in oceanic layer 2 resulting from a preferred alignment of fractures has been widely recognized, but all experiments to date have sought to measure only the weak azimuthal variation of elastic properties resulting from tectonically controlled systems of vertical fractures. From ocean drilling data, however, especially from DSDP Hole 504B, we know that layer 2 is composed of interleaved massive flows and breccia units, and that the massive units have a very strong concentration of horizontal fractures. Layer 2's pronounced horizontal fabric of low-velocity "layers" (fractures and/or breccia zones) permeating an otherwise high-velocity matrix, will cause *P*-waves to travel faster horizontally than vertically. This anisotropy has no azimuthal expression, and so cannot easily be recognized in seismic data, but it may lead to overestimation of the thickness of upper crustal layers by as much as 30% in young crust. Further, the anisotropy affects *P* and *S* waves differently, so where shear-wave data are available, Poisson's ratio may be substantially underestimated. The widespread observation of a low Poisson's ratio zone in the upper few kilometers of young crust is almost certainly an artifact of ignoring anisotropy. As the crust ages, fractures and voids are filled by chemical alteration and precipitation, the velocity contrast between rock and void-filling material is reduced, and the anisotropy decreases. The errors introduced by assuming isotropy thus show an inverse relationship to crustal age, so that thickness measurements from old crust are probably no more than 10% in error. This explains a long-standing enigma of marine seismology: the apparent thinning of upper crustal layers with age.

Introduction

In a classic paper describing a compilation of a very large quantity of sonobuoy refraction data, Houtz and Ewing [1976] revealed two systematic age-dependent properties of the uppermost oceanic crust, layer 2A: its seismic velocities increase with age while its thickness decreases. Since that study, improved seismic data have demanded substantial modification of Houtz and Ewing's crude layer-cake interpretations [Spudich and Orcutt, 1980b], but the systematic increase of layer 2A velocities with age has been confirmed. In young crust layer 2A has low velocities but extremely strong velocity gradients [Bratt and Purdy, 1984; Purdy, 1987], while in older crust it has

higher velocities and lower gradients [Purdy, 1983]. Lateral variability of structure may obscure this trend locally [Stephen, 1988], but the great body of marine seismic data, taken together, clearly shows that mean velocities follow Houtz and Ewing's trends.

The thinning of layer 2A is less well resolved. While Houtz and Ewing's statistics show clearly that 2A thins with age [Houtz and Ewing, 1976], subsequent seismic refraction surveys have not identified any age-dependent thickness on a regional scale. The failure of these more modern experiments, however, may result from investigation of too small an age range or proximity to confusing tectonics such as fracture zones. Only one series of refraction experiments has covered the range 0–10 M.y. in a region free of fracture zones, that of Bunch and Kennett [1980] on the Reykjanes Ridge. Bunch and Kennett saw not a thinning of layer 2A, but an equally puzzling thinning of the more constant velocity layer beneath, layer 2B.

The filling of fractures and pores in the shallow crust by alteration products and hydrothermal mineralization is generally believed to explain the increases in velocity with age and to be largely responsible for velocity gradients [Schreiber and Fox, 1976, 1977; Bratt and Purdy, 1984]. If such void filling makes the velocity age dependent, then it seems probable that it also controls the age dependence of the thickness of upper crustal layers. The obvious explanation for the thinning of layer 2A, that velocities increase through healing of voids until the layer cannot be resolved seismically from layer 2B, demands that layer 2B thicken while layer 2A thins. No such thickening is seen in seismic data. Indeed, on the Reykjanes Ridge, layer 2B is seen to thin by 500 m over 9 M.y. while the gradient zones of layers 2A and 2C remain fairly constant in thickness [Bunch and Kennett, 1980]. Clearly, a thinning layer underlain by a layer of constant thickness is incompatible with the idea of one layer growing at the expense of the other through hydrothermal deposition. It seems reasonable to assume that the rate at which crustal layers have been produced has not increased with time, but this leaves the observed thinning of upper crustal layers as an enigma.

Another enigma of marine seismology, apparently unrelated to the problem of thinning layers, is the seismic refraction observation of very low values of Poisson's ratio at depths of 1–3 km into the crust [Spudich and Orcutt, 1980a; Au and Clowes, 1984; Chiang and Detrick, 1985]. On the basis of laboratory measurements of ophiolite samples [Salisbury and Christensen, 1978; Christensen and Smewing, 1981], it is difficult to justify a Poisson's ratio of less than 0.26 anywhere in the crust, yet refraction data demand minima as low as 0.23. This disagreement is particularly troubling as it seems to represent a failure of the widely-held idea that ophiolites represent obducted sections of oceanic crust. Two explanations have been presented: (1) that the Poisson's ratio minima result from trondhjemites at much shallower levels and in much greater volume than they occur in ophiolites [Spudich and Orcutt, 1980a; Chiang and Detrick, 1985] (this

---
[1]now with the Department of Geological Sciences, AJ-20, University of Washington, Seattle, WA 98195

explanation itself violates the ophiolite analogue), and (2) that the minima are caused by much thicker cracks and much higher porosity than seems reasonable for these depths [Shearer, 1988]. Neither explanation is particularly convincing, and their authors seem quite willing to accept alternate explanations. The low values of Poisson's ratio have remained a puzzle.

We show here that the thinning of crustal layers, and the minima in Poisson's ratio, are apparent, not real. These anomalies are explained by the fabric of layer 2. Not only does the layering of extrusives impose a strong horizontal fabric on layer 2, but the geometry of fractures is critically important also. It is well accepted that the uppermost oceanic crust must be profoundly fractured, since only with large-scale porosity is it possible to reconcile P velocities of $5.9\,\mathrm{km\,s^{-1}}$ measured in drill samples of basement rocks [Hyndman and Drury, 1976] with seismic refraction observations of velocities as low as $2.1\,\mathrm{km\,s^{-1}}$ [Whitmarsh, 1978; Spudich and Orcutt, 1980a; Purdy, 1987]. Even with very high porosity it may be necessary to invoke pore pressure effects to explain such low seismic velocities [Spudich and Orcutt, 1980a; Christensen, 1984]. What has not been recognized is that fracture orientation too profoundly affects the seismic properties of the shallowest igneous crust.

That there are fractures with a preferred orientation in the oceanic crust is dramatically evident from any sidescan sonar image of a mid-ocean ridge [e.g., Fornari et al., 1988]. Such large-scale fractures are near vertical and introduce an azimuthal variation in seismic properties. Such fracture-induced azimuthal anisotropy was first detected by Stephen [1981], whose findings have since been extensively corroborated [White and Whitmarsh, 1984; Shearer and Orcutt, 1985; Stephen, 1985]. That another fracture set might coexist with the vertical fractures was completely unsuspected until Newmark et al. [1985] reported predominantly horizontal fractures occurring throughout layer 2 at DSDP site 504B (the fractures are indeed horizontal; Newmark et al. [1985] corrected for the fact that a vertical drill hole emphasizes horizontal features). Combined with the fabric imposed by flow layering, the horizontal fractures must induce a very strong seismic anisotropy such that compressional-wave energy travels faster in a horizontal direction than it does vertically. Standard refraction experiments cannot detect such anisotropy, however, so when upper crustal anisotropy is considered at all, it is invariably limited to the effects of vertical fractures. A natural consequence of this omission is that the thickness of crustal layers is systematically overestimated and Poisson's ratio underestimated. The anisotropy should decay with age, however, since it is primarily caused by voids and fractures which will fill as alteration and hydrothermal deposition continue. The errors in thickness should consequently also decay with age, making crustal layers appear to thin, while at the same time Poisson's ratio anomalies should appear to fall. Such behaviour is precisely what is observed.

Notation.

The notation of anisotropy is confusing so ambiguous terms abound in the literature. To keep things as intuitive as possible, we follow the recommendation of Crampin [1989] and use the term "azimuthal isotropy" to describe anisotropic systems which display no azimuthal variation of properties (the widely-used term "transverse isotropy," meaning isotropic transverse to some symmetry axis, is synonymous only when the symmetry axis is vertical). We find it helpful to define another term, "vertical anisotropy," meaning an anisotropic system in which there is variation of properties in any vertical plane; a vertically anisotropic medium need not be azimuthally isotropic (for example, an orthorhombic system is both vertically and azimuthally anisotropic).

Fractures, Fabric, and Velocity Variation

Layer 2 has a strongly-developed horizontal fabric, since it is made up of horizontally-lying thin flows, pillow units, massive lavas, and breccia zones, with intercalated sediment. This fabric is further enhanced by fractures: Newmark et al. [1985] report from televiewer (acoustical) images of the walls of Hole 504B that the pillows and massive units are extensively fractured, with the great predominance of fractures subhorizontal. A fabric displaying any preferred direction will render a medium anisotropic. We show the simplest possible fabrics in Figure 1, which shows media made up of high velocity particles imbedded in a low-velocity matrix. In Figure 1a the medium is random, displaying no preferred direction. A compressional wave impinging on a random medium "sees" the same bulk properties regardless of its direction, so its velocity is independent of direction. In contrast, a horizontal fabric (Figure 1b), will cause velocity to vary with angle of incidence. A P-wave travelling horizontally through the material of Figure 1b preferentially propagates through the higher velocity material, but a vertically travelling wave cannot avoid the lower velocity material, and so must travel at a lower "average" velocity. If the only fabric is horizontal, then velocity varies only with angle of incidence, and the medium is azimuthally isotropic.

Representative velocity variations in a vertical plane for isotropic and anisotropic media are shown in Figure 2. The exact form of the velocity variation for the anisotropic medium will depend on its elastic parameters; in general, azimuthal isotropy demands five elastic parameters to describe the velocity variation while isotropy requires only two. Note from Figure 2 that while the P-velocity is faster horizontally than vertically, the SV-velocity is the same in these directions. Note also that shear-wave triplications may occur (i.e., in certain directions, three SV arrivals may be detected).

A horizontal fabric results in vertical anisotropy, but the cause of that fabric is immaterial to the seismic response. Aligned fractures or pores, preferred orientation of grains, or any type of small-scale layering, will all result in anisotropy; these different causes will be seismically indistinguishable as long as the scale of the fabric is small compared to a seismic wavelength [Backus, 1962]. In seismic measurements of upper oceanic crust, the wavelength of a compressional wave is invariably greater than 50 m. This greatly exceeds the typical thickness of individual pillow, breccia, flow, or massive units: at DSDP Hole 504B, each unit is typically less than 5 m thick [Cann, Langseth, Honnorez, von Herzen, White, et al., 1983; Anderson, Honnorez, Becker, et al., 1985]. Although units as much as 50 m thick are observed,

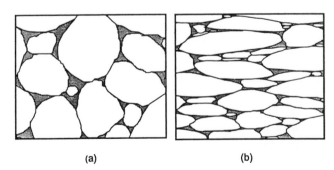

(a)   (b)

Fig. 1. A random medium (a) and the same medium foreshortened vertically so that it has a horizontal fabric (b). The particles have one velocity and the infilling matrix another. In (a), seismic velocities are independent of direction, but in (b), P waves will travel faster horizontally than vertically.

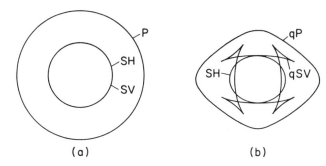

Fig. 2. Vertical sections of expanding wavefronts in (a) an isotropic medium and (b) a vertically anisotropic medium with azimuthal isotropy (the vertical axis is an axis of symmetry for both figures). In the isotropic medium both P and S wavespeeds are independent of direction, and the shear waves are degenerate, with SV and SH having the same speed. In the anisotropic medium, P and SH are faster horizontally than vertically while the speed of SV is the same in the horizontal and vertical directions. The SV wavefront may display triplications.

these are either subdivided into thinner subunits (such as individual pillows in a pillow lava sequence) or are moderately to severely disrupted by horizontal fractures [Newmark, et al., 1985], so that the entire thickness of layer 2 has a well-developed horizontal fabric with a characteristic scale length of a few meters. Other deep drill holes reveal similar fabric [e.g., Melson, Rabinowitz, et al., 1979; Donnelly, Francheteau, Bryan, Robinson, Flower, Salisbury, et al., 1980], but only 504B samples the deeper part of layer 2. It is clearly risky to extrapolate to the entire oceanic crust from a single drill hole, but since flows and pillows in ophiolites have similar dimensions to those inferred from site 504B [e.g., Casey, et al., 1981; Christensen and Smewing, 1981], we conclude that layer 2 does indeed have a well-developed horizontal fabric everywhere. Layer 2 must necessarily be vertically anisotropic.

Previous studies have not seriously considered the possibility of layer 2 anisotropy other than the azimuthal variability imposed by vertical fractures. Whitmarsh [1978] casually noted that anisotropy might result from the multiple layering of flows, and Newmark et al. [1985] observed that the horizontal fractures seen in televiewer data should make the medium strongly anisotropic, but the matter has not been pursued because no vertical anisotropy has been observed in seismic data.

That vertical anisotropy has not been identified in layer 2 demands some explanation. Vertical anisotropy has been widely identified in marine sediments [Bachman, 1979], principally from laboratory measurements. The in situ identification of such anisotropy has only been possible in sedimentary formations where the same identifiable horizon exists in both reflection and refraction data [e.g., Davis and Clowes, 1986], or when depths computed from reflection data disagree with depths obtained directly by drilling [e.g., Banik, 1984]. In the case of layer 2, the sample size for laboratory measurement of velocity from core samples is too small to sample the seismic effects of fractures or fabric, while the lack of a uniform prominent reflecting/refracting horizon in the upper crust means that a mismatch between reflection and refraction results, indicative of vertical anisotropy, has never been noticed. By contrast, anisotropy induced by vertical fractures which has been widely observed in layer 2 [Stephen, 1981, 1985; White and Whitmarsh, 1984; Shearer and Orcutt, 1985]. Vertical fracturing gives rise to an azimuthal anisotropy which can be

detected seismically by shooting a circle around a receiver and mapping the variation of properties with direction. Vertical anisotropy arising from a horizontal fabric displays no such azimuthal variation.

Although it has not been observed, the vertical anisotropy is probably very large. The very high fracture concentrations reported by Newmark et al. [1985], together with the high contrast in seismic velocity between unfractured basalt and fracture-filling water, imply a very strong anisotropy dominated by fracture-induced effects. Extrapolating from the results of Anderson et al. [1974] it seems likely that in areas where there is little hydrothermal deposition the fracture-induced anisotropy approaches 40% for P-waves.

Velocity Gradients and Vertical Anisotropy

Where vertical anisotropy occurs in a region of velocity increase with depth, the effects of the two phenomena, anisotropy and velocity gradient, are intrinsically coupled; one cannot be considered at the exclusion of the other. In the oceanic crust, velocity gradients are greatest in the upper two kilometers, in the same depth range where fractures and fabric suggest appreciable anisotropy. Unfortunately, marine seismic experiments sample this part of the crust inadequately. Our knowledge of seismic structure is gleaned almost entirely from sparse, poor quality P-refraction data. These data can always be explained adequately without invoking anisotropy, but the final models resulting from such analysis, while self-consistent, may be profoundly wrong, as we shall show.

We illustrate the problem schematically in Figure 3. A horizontally-fractured ocean crust, in which velocities and aligned fracture concentration both increase with depth, has a velocity-depth dependence something like the heavy curves in Figure 3c. At any given depth, the velocity of horizontally travelling P waves, v_{PH}, is higher than that of vertically travelling waves, v_{PV}. Regardless of the degree of

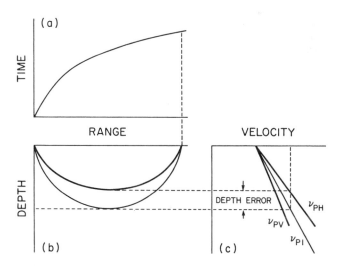

Fig. 3. Errors in depth determinations.
(a) P-refraction travel time plot for source and receiver on a horizontally fractured ocean bottom with a velocity gradient.
(b) The true raypath through the anisotropic medium (heavy curve) and the raypath assumed by any isotropic inversion procedure (light curve).
(c) The true velocity-depth function (heavy lines); v_{PH} is horizontal velocity, v_{PV} is vertical velocity. The inversion procedure picks the light line labelled v_{PI}.

fracture-induced anisotropy, from P-refraction data alone it will be impossible to detect any anisotropy at all. The travel-time curve for P-refractions, Figure 3a, gives no indication of anisotropy, but in an actual experiment, this travel-time curve will comprise the only information available. Invariably, isotropy will be assumed. Inversion to obtain a velocity-depth function proceeds by determining the horizontal velocity at a turning point (from the slope of the travel-time curve of Figure 3a), then adjusting depths to satisfy travel times. With no indication of anisotropy, we can only assume that velocities are the same in all directions. If, as in this case, vertical P velocities are really less than horizontal, then a ray will not penetrate to as great a depth to yield the same travel time. A refracted ray will have a shallower turning point than the ray implicitly assumed by the inversion scheme, as shown in Figure 3b. Figure 3c shows the consequences of this depth difference for the inversion. The velocity-depth function deduced from travel-time analysis, v_{PI}, lies somewhere between the true vertical and horizontal velocities. Since the v_{PI} curve always lies below the v_{PH} curve, seismic refraction data will systematically overestimate depths.

Age Dependence

We have suggested above that the apparent thinning of crustal layers with age results from a reduction in anisotropy. To investigate this we have considered a series of upper crustal models loosely based on the idealized structure of Bratt and Purdy [1984], modified to include the anisotropic effects of horizontal fractures and varying degrees of fracture filling. The objective here was to see how realistic degrees of anisotropy may have contributed to misinterpretation of seismic data.

The first model, representing young crust ("Age Zero"), is shown by the first of the diagrams in Figure 4. This model has 8% anisotropy at the top of layer 2A, increasing to 35% at the base of layer 2A, at a depth of 500 m below the seafloor. Such increase in anisotropy with depth is broadly consistent with the observation that fracture density increases with depth in layer 2A [Newmark et al., 1985], but such large anisotropy would only be possible if the cracks were predominantly water-filled. In layer 2B of the model (500–1000 m depth), velocities and anisotropy increase more slowly with depth. In layer 2C (1000–1500 m depth) the anisotropy decreases, consistent with the idea that fractured flows progressively give way to unfractured dikes as depth increases. Throughout the entire model, elastic parameters were chosen to be consistent with a horizontally-fractured medium, following the theory of Schoenberg [1983].

We computed P travel times for this vertically anisotropic structure and then inverted them, assuming isotropy, using the τ-sum method of Diebold and Stoffa [1981]. The line labelled v_{PI} in Figure 4 shows the inversion results. Because of the depth errors intrinsic to the assumption of isotropy, layer 2A thickness has been overestimated by over 100 m and overall layer 2 thickness by almost 400 m. Ignoring the anisotropy has introduced errors of about 25%.

As the crust ages, alteration products and hydrothermal deposits fill the cracks, with mineralization growing inward from the walls of the crack. This reduces crack dimensions and so increases the overall seismic velocity. The mineralization also changes the crack aspect ratios: as the deposits on the walls grow, cracks and fractures will get proportionately thinner (the aspect ratios will decrease), until bridging between the walls occurs to produce isolated void spaces each with larger aspect ratio (i.e., closer to spherical) than the original fracture. With both crack thinning and bridging going on, it is not at all clear how the mean aspect ratio will vary. Although anisotropy is sensitive to aspect ratio [Shearer, 1988], we expect that any anisotropy change related to a change in mean aspect ratio will be much smaller than the decrease in anisotropy resulting from the reduction in velocity contrast between rock and crack-filling material. Overall, velocities will increase while gradients and anisotropy decrease.

To investigate the consequences of these time dependent effects, we have assumed a linear change in the elastic stiffness with "time," and performed the same ray-tracing and inversion exercise described above for "Age Zero" for a variety of older models, also shown in Figure 4: "Age Five" has a maximum anisotropy of 20%, and "Age Ten" a maximum anisotropy of 8%. As expected, as the anisotropy decreases, the depth errors introduced by assuming isotropic structure also decrease.

Since the depth errors decrease as a function of the amount of fracture filling, crustal layers will appear to thin with age. Figure 5 demonstrates this result for all of layer 2. The apparent thinning is completely consistent with Bunch and Kennett's [1980] measurements on the Reykjanes Ridge. Figure 6 shows the apparent thinning just for layer 2A; the curve is remarkably similar to Houtz and Ewing's [1976] results. While Houtz and Ewing [1976] suggested that layer 2A thins to zero thickness, the resolution of their data was probably only about 500 m, so the apparent thinning we predict quite adequately explains their observations.

Fig. 4. Velocity-depth functions for young (Age Zero), intermediate (Age Five) and old (Age Ten) oceanic crust. Layers 2A, 2B and 2C comprise the depth range 0–500 m, 500–1000 m and 1000–1500 m respectively. Two curves define the P velocity in each case: v_{PV} for vertical propagation and v_{PH} for horizontal propagation. True shear velocity, v_{SV}, is marked by a single curve, as both vertically and horizontally travelling SV waves have the same speed. Gradients are high and anisotropy (apparent from the separation of the v_P curves) is initially low in Layer 2A. Fracture concentration increases through 2A so anisotropy increases. Layer 2B has a much lower gradient and near-uniform anisotropy. Fracture concentration (and hence anisotropy) falls through Layer 2C and is zero in Layer 3. Curves v_{PI} and v_{SI} show the inversions which would be obtained if the structures were assumed to be isotropic and P and S travel times were inverted separately. Note that the P and S inversions disagree in depth. The errors in the P-inversions decrease with age as the anisotropy decreases due to fracture filling.

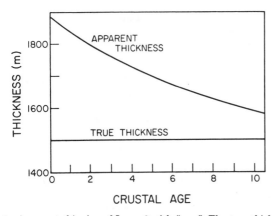

Fig. 5. Apparent thinning of Layer 2 with "age." The true thickness is constant but the layer appears to thin because the errors introduced by anisotropy decrease as the anisotropy decreases.

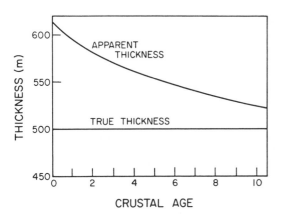

Fig. 6. Apparent thinning of Layer 2A with "age." If the resolution of seismic data were poorer than 500 m, the layer would appear to thin to zero thickness, as suggested by Houtz and Ewing [1976].

Shear Waves and Poisson's Ratio

Shear-wave velocity variations in an azimuthally isotropic medium are quite different from the compressional variation, as we have seen in Figure 2. An important point to note from Figure 2 is that vertically and horizontally propagating SV waves have the same velocity (even though the velocity may be different in some other direction). Consequently, inversions of shear-wave refraction data produce much smaller depth errors than those suffered in a P inversion. This can be seen in Figure 4 by comparing the v_{PI} and v_{SI} curves with the true structure.

Since the P and S depth errors differ, if isotropic inversions are believed, an S velocity from one depth will be associated with a P velocity which actually belongs to some shallower depth; the P velocity will seem to be too small. Any quantity depending on both P and S velocities will display anomalous behaviour. One such quantity is Poisson's ratio σ, which, for an isotropic material, is given by

$$\sigma = \frac{1}{2}\left[\frac{v_P^2 - 2v_S^2}{v_P^2 - v_S^2}\right],$$

where v_P and v_S are P and S velocities. Since the P velocity is too small for the S velocity, Poisson's ratio will be systematically underestimated. This is illustrated in Figure 7, which shows both the true Poisson's ratio and the apparent Poisson's ratios deduced from the isotropic inversions of Figure 4.

It is apparent that if the upper crust has vertical anisotropy, we should expect to find anomalously low Poisson's ratios when we interpret seismic refraction data assuming isotropy. Such anomalies are widely observed. Figure 8 shows the apparent Poisson's ratio trajectory of Spudich and Orcutt [1980a] for 15 M.y.-old crust. The curve shows a distinct minimum of $\sigma = 0.24$ at a depth of about 1.4 km. Au and Clowes [1984] and Chiang and Detrick [1985] have found similar Poisson's ratio minima. In all of these cases, the true cause of the Poisson's ratio minimum is probably a misinterpretation of the seismic data. The assumption of isotropy forces us to compare P and S velocities from different depths and a low Poisson's ratio is a necessary consequence. We have made no attempt to model the Spudich and Orcutt [1980a] result shown in Figure 8, yet its similarity to the Age Five curve of Figure 7 is so remarkable that we are confident enough to assert that the crust at Spudich and Orcutt's site contains partially-filled horizontal fractures down to 1.5 km depth.

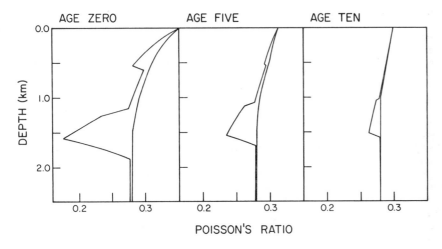

Fig. 7. Poisson's ratio as a function of depth for the structures of Figure 4. The heavy curves show the correct (horizontal) Poisson's ratios, the light curves show the Poisson's ratios that would be inferred from the isotropic inversions. Note that the apparent Poisson's ratio is systematically too low, but that the error decreases with age.

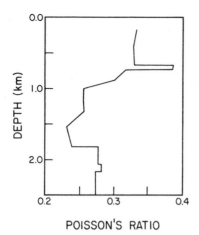

Fig. 8. Apparent Poisson's ratio curve from Spudich and Orcutt [1980a]. Poisson's ratio appears to reach a minimum of about 0.24 at about 1.5 km depth. Compare this minimum with that for the Age Five structure in Figure 7.

As final evidence that vertical anisotropy explains the Poisson's ratio minima we consider Shearer and Orcutt's [1986] refraction results from 140 M.y.-old lithosphere in the Western Pacific. These authors measured azimuthal anisotropy in the upper crust, almost certainly the result of vertical fractures, but they did not consider the possibility of any superposed vertical anisotropy caused by horizontal fractures. Nevertheless, convincing evidence for such anisotropy exists within their published results. After a very careful travel-time analysis, Shearer and Orcutt were unable to reconcile the P and S-wave data: their S-wave profile reaches layer 3 velocities at 1.7 km depth versus 2 km for the P-velocity profile. This is a classic case of overestimation of depth from P-wave data indicative of vertical anisotropy. If we combine Shearer and Orcutt's P and S profiles we obtain a Poisson's ratio minimum of about 0.26 at a depth of 1.7 km, very similar to but slightly deeper than the anomaly for the Age Ten crust in Figure 4. We conclude that the crust at Shearer and Orcutt's site is horizontally fractured, but that the fractures are substantially more filled than at Spudich and Orcutt's [1980a] 15 M.y.-old site. It is worth noting that if we demand that the layer 2-layer 3 boundary be at the same depth for both P and S waves, then Shearer and Orcutt's data are incompatible with a vertically isotropic upper crust.

Discussion

We have argued that alteration and hydrothermal deposition will change the geometry of fractures and other voids through time and so modify the seismic structure of the upper crust. While age will be the controlling factor in determining the degree of healing of cracks through low-temperature alteration [Schreiber and Fox, 1977], the effects of hydrothermal deposition will depend more on the distribution and vigor of hydrothermal circulation cells than on age per se. We should expect that in areas of vigorous hydrothermal activity the modification of crack volume and aspect ratio by mineralization will result in low anisotropy and increased seismic velocities, regardless of age. Since hydrothermal circulation must vary both along and across strike, we should expect lateral heterogeneity in the upper crust. Such heterogeneity is indeed observed [Stephen, 1988]. The arbitrary "age" of Figures 4–7 is thus in part really some cumulative measure of hydrothermal deposition.

The fact that there is both apparent thinning and a reduction in the Poisson's ratio anomaly with age confirms what we have already suggested, that the vertical anisotropy arises primarily from the horizontal fractures rather than from the fabric of superposed flows. Anisotropy resulting purely from the fabric of the layering must be independent of age, while fracture-induced anisotropy will show an age dependence because of the time-dependent void geometry. Even completely filled fractures would produce some residual anisotropy, because of the velocity contrast between fracture-filling material and the unfractured basalt. The oceanic crust probably never attains complete fracture filling, however: even in old crust, porosity is high [Purdy, 1987], so it is probable that fractures too are only partially filled. Very crude estimates based on the anisotropy implied by Shearer and Orcutt's [1986] structure (the one datum we have from old crust) suggest that the fractures are only half-filled by 140 M.y. age. Since the partial filling of fractures yields such a clear age dependence, we suspect that fractures produce almost all of the anisotropy, and that the effects of non-fracture fabric are negligible.

We have not yet discussed the origins of the fracturing. Newmark et al. [1985] suggest two possible causes for fractures at Hole 504B: natural hydrofracturing of wet rock during dike injection, and mechanical rebound with brittle failure immediately following the intrusion of a dike into rock under high tension. We regard natural hydrofracturing as the more plausible mechanism. The argument here is that thermal pressurization of water during dike injection can be adequate to cause the host rock to fail, especially the more massive and less porous units. Newmark et al. [1985] applied Hole 504B physical properties to the theory of Delaney [1982] and determined that horizontal stresses would be greater than vertical stresses, explaining the very shallow dip of the fractures. If that is indeed the mechanism, then the same phenomenon probably occurs at all spreading centers, making horizontal fractures a ubiquitous feature of shallow ocean crust.

While we believe that we have presented here a strong case for vertical anisotropy in the shallow crust, it is important to note that we have not yet performed an actual inversion of seismic data to show that such anisotropy must exist. We have simply hypothesized particular styles of anisotropy consistent with televiewer observations and showed that vertical anisotropy provides more plausible explanations for seismic refraction data than more traditional isotropic models. The best technique to measure this anisotropy would be to use downhole seismometers and vertical seismic profiling, but this clearly could only be done at a drill hole and then would only provide measurements at a single point. For studies of age dependence or along-axis variation multiple drill holes would be required, a prohibitive expense. Unfortunately, the direct measurement of vertical anisotropy from traditional ocean-bottom seismometer refraction data is not possible, as the inverse problem is underconstrained if only P and SV refractions are recorded. With additional information, however, either P and SV reflections from some horizon beneath the anisotropic layers, or SH refractions, or preferably both, stable inversion is possible [Miller et al., 1987]. Shear wave information is critical, and for this reason the use of ocean-bottom seismometers in marine refraction experiments should be encouraged. The value of SH information means that serious consideration should be given to the development of on-bottom shear wave sources capable of putting enough energy into the bottom to sample depths of up to two kilometers.

A single experiment would provide a measure of the anisotropy at a point, but determining the age dependence requires a series of experiments. Because of the along-axis variability of spreading style, measurements must be made along a single seafloor spreading flow line. The global age relationships described by Houtz and Ewing [1976], layer thinning and velocity increases, were not apparent in measurements made on 0.5 and 4.0 M.y.-old crust in the ROSE area [Purdy, 1982], but those measurements were taken on crust formed at different segments of a rise axis. By contrast, using measurements at 0, 3 and 9 M.y. age along a single flow line on the Reykjanes Ridge, Bunch and Kennett [1980] were able to confirm the Houtz and Ewing trends. Following the same sampling philosophy, to investigate the age dependence of the anisotropy, we recommend small-scale on-bottom experiments at three or more locations on 0-to-10 M.y.-old crust along a single flow line.

We have discussed here only vertical anisotropy resulting from horizontal layering and fractures. Since the vertical fractures in the crust also induce an anisotropy, the actual anisotropy of the crust must have contribution from both systems. The velocity variation resulting from vertical fractures alone has a hexagonal symmetry with a horizontal symmetry axis aligned perpendicular to the fractures [Crampin, 1978; Schoenberg, 1983]. In seismic experiments the symmetry axis is typically found to be parallel to the spreading direction [e.g., Stephen, 1985], implying that the vertical fractures are perpendicular to spreading, as expected from tectonics. With such symmetry there will be no anisotropy perpendicular to spreading, but there will be anisotropy in all other vertical planes. Most of the refraction lines we have considered here were perpendicular to spreading, so vertical fractures cannot explain the vertical anisotropy we have detected. The fact that vertical fractures will give rise to a vertical anisotropy along other azimuths, however, emphasizes the need to know the alignment of refraction lines relative to local tectonic features. It is clear that any quantitative investigation of anisotropy in the oceanic crust must admit the possibility of both vertical and horizontal fracture systems and so should measure the full three-dimensional variation of velocity with direction.

Conclusions

Horizontal fractures within layer 2 render the layer anisotropic, but the anisotropy has no azimuthal expression and so cannot easily be detected seismically. A simple assessment of the effects of this anisotropy shows that two major enigmas of marine seismology, the thinning of upper crustal layers with age and the existence of a zone of anomalously low Poisson's ratio, are really artifacts resulting from the erroneous assumption that the crust is isotropic. Expanding seismic wavefronts are distorted by the anisotropy; in particular, P wavefronts are distorted such that layers appear to be thicker than they really are. The apparent thinning of layers indicates a decay of anisotropy with age (and hence a reduction in wavefront distortion) resulting from the filling of the fractures by alteration products and hydrothermal precipitation. This suggests that studies of the anisotropy should yield information on the nature and longevity of hydrothermal systems.

The wavefront distortion resulting from the fractures is different for P and S waves, so the combination of P and S data to get Poisson's ratio yields erroneous values which are systematically too small. Low Poisson's ratios are never seen in ophiolites, so the demonstration that the Poisson's ratio anomalies are artifacts is particularly important: the major inconsistency between the ophiolite model of the ocean crust and seismic refraction data has been removed.

The only direct evidence we have for the horizontal fracturing is from the borehole televiewer imagery of Newmark et al. [1985]. There have been other televiewer lowerings in ocean drill holes, but all other published images have been degraded by ship motion. New digitally recording televiewers are now available which permit post-log image reconstruction. Use of these new instruments, and display of the resulting data as travel-time images as well as the traditional amplitude images, should greatly enhance the identification of fractures. The use of such devices (with heave compensation if seas are too large) should be encouraged in every drill hole penetrating more than 50 m into igneous crust.

We do not yet know how to measure the fracture-induced anisotropy directly, although it is clear that a successful experiment will have to use on-bottom sources and ocean bottom (or downhole) seismometers. In the meantime, it is essential that the possibility of anisotropy be kept in mind during analysis of seismic refraction data from the oceans. While the anisotropy resulting from vertical fractures can be avoided by orienting refraction lines perpendicular to the spreading direction, it is impossible to escape the anisotropic effects of horizontal fractures. Even when this vertical anisotropy is small, the assumption of isotropy can lead to significant errors in depth. To avoid such errors, inversion of seismic data should be guided by comparing field seismograms to synthetics which include the effects of azimuthal isotropy (computed, for example, following the techniques of Fryer and Frazer [1987]). If such synthetic seismogram software is unavailable, analysing P and S data separately will at least preserve the differences in structure from which a departure from isotropy can be estimated. The most convincing of the evidence presented here for this anisotropy are the depth disagreements between P and S data made evident in the Poisson's ratio minima of Spudich and Orcutt [1980a], Au and Clowes [1984], and Shearer and Orcutt [1986]. Driven by a desire to match their data as accurately as possible, these authors resisted the temptation to force similar features in their P and S profiles to occur at the same depths; from the curious structures resulting from their analyses, vertical anisotropy can confidently be inferred. It is likely, however, that the interpretation of data from other refraction experiments, especially in older crust, has implicitly assumed isotropy by forcing all interfaces in the structure to be at the same depth in both P and S profiles. It seems probable that there is much to learn about anisotropy, (and, by inference, about hydrothermal systems and crustal ageing), from seismic data already in existence.

Acknowledgments. This research was supported by the Office of Naval Research and from National Science Foundation grants OCE-8516225 and OCE-8711646. Hawaii Institute of Geophysics contribution no. 2095.

References

Anderson, D. L., B. Minster, and D. Cole, The effects of oriented cracks on seismic velocities, *J. Geophys. Res., 79,* 4011–4015, 1974.

Anderson, R. N., J. Honnorez, K. Becker, et al., Hole 504B, Leg 83, *Initial Rep. Deep Sea Drill. Proj., 83,* 13–118, 1985.

Anderson, R. N., H. O'Malley, and R. L. Newmark, Use of geophysical logs for quantitative determination of fracturing, alteration, and lithostratigraphy in the upper oceanic crust, Deep Sea Drilling Project, Holes 504B and 556, *Initial Rep. Deep Sea Drill. Proj., 83,* 443–478, 1985.

Au, D., and R. M. Clowes, Shear-wave velocity structure of the oceanic lithosphere from ocean bottom seismometer studies, *Geophys. J. R. Astron. Soc., 77,* 105–123, 1984.

Bachman, R. T., Acoustic anisotropy in marine sediments and sedimentary rocks, *J. Geophys. Res., 84,* 7661–7663, 1979.

Backus, G. E., Long-wave elastic anisotropy produced by horizontal layering, *J. Geophys. Res., 67,* 4427–4440, 1962.

Banik, N. C., Velocity anisotropy of shales and depth estimation in the North Sea basin, *Geophysics, 49,* 1411–1419, 1984.

Bratt, S. R., and G. M. Purdy, Structure and variability of oceanic crust on the flanks of the East Pacific Rise between 11° and 13°N, *J. Geophys. Res., 89,* 6111–6125, 1984.

Bunch, A. W. H., and B. L. N. Kennett, The crustal structure of the Reykjanes Ridge at 59°30′N, *Geophys. J. R. Astron. Soc., 61,* 141–166, 1980.

Cann, J. R., M. G. Langseth, J. Honnorez, R. P. Von Herzen, S. M.

White, et al., Sites 501 and 504: sediments and ocean crust in an area of high heat flow on the southern flank of the Costa Rica Rift, *Initial Rep. Deep Sea Drill. Proj., 69,* 31–173, 1983.

Casey, J. F., J. F. Dewey, P. J. Fox, J. A. Karson, and E. Rosencrantz, Heterogeneous nature of oceanic crust and upper mantle: a perspective from the Bay of Islands ophiolite complex, in *The Sea,* vol. 7, *The Oceanic Lithosphere,* edited by C. Emiliani, pp. 305–338, J. Wiley & Sons, New York, 1981.

Christensen, N. I., Pore pressure and oceanic crustal seismic structure, *Geophys. J. R. Astron. Soc., 79,* 411–423, 1984.

Christensen, N. I., and J. D. Smewing, Geology and seismic structure of the northern section of the Oman ophiolite, *J. Geophys. Res., 86,* 2545–2555, 1981.

Chiang, C. S., and R. S. Detrick, The structure of the lower oceanic crust from synthetic seismogram modeling of near-vertical and wide-angle reflections and refractions near DSDP site 417 in the Western North Atlantic (abstract), *Eos, Trans. Am. Geophys. Union, 66,* 956, 1985.

Crampin, S., Seismic wave propagation through a cracked solid: polarization as a possible dilatancy diagnostic, *Geophys. J. R. Astron. Soc., 53,* 467–496, 1978.

Crampin, S., Suggestions for a consistent notation for seismic anisotropy, *Geophys. Prospecting,* in press, 1989.

Davis, E. E., and R. M. Clowes, High velocities and seismic anisotropy in Pleistocene turbidites off Western Canada, *Geophys. J. R. Astron. Soc., 84,* 381–399, 1986.

Delaney, P. T., Rapid intrusion of magma into wet rock: groundwater flow due to pore pressure increases, *J. Geophys. Res., 87,* 7739–7756, 1982.

Diebold, J. B., and P. L. Stoffa, The traveltime equation, tau-p mapping, and inversion of common midpoint data, *Geophysics, 46,* 238–254, 1981.

Donnelly, T., J. Francheteau, W. Bryan, P. Robinson, M. Flower, M. Salisbury, et al., Site 418, *Initial Rep. Deep Sea Drill. Proj., 51-53, Part 1,* 351–626, 1980.

Fryer, G. J., and L. N. Frazer, Seismic waves in stratified anisotropic media – II. Elastodynamic eigensolutions for some anisotropic systems, *Geophys. J. R. Astron. Soc., 91,* 73–101, 1987.

Fornari, D. J., J. A. Madsen, D. G. Gallo, M. R. Perfit, and A. N. Shor, Morphology of the East Pacific Rise from the axis to seafloor ~4 ma old between 13°–15° N: Bathymetric anomalies and local variations in subsidence along a ridge segment (abstract), *Eos, Trans. Am. Geophys. Union, 69,* 1485, 1988.

Houtz, R., and J. Ewing, Upper crustal structure as a function of plate age, *J. Geophys. Res., 81,* 2490–2498, 1976.

Hyndman, R. D., and M. J. Drury, The physical properties of oceanic basement rocks from deep drilling on the Mid-Atlantic Ridge, *J. Geophys. Res., 81,* 4042–4052, 1976.

Melson, W. G., P. D. Rabinowitz, et al., Site 395: 23°N Mid-Atlantic Ridge, *Initial Rep. Deep Sea Drill. Proj., 45,* 131–264, 1979.

Miller, D. J., G. J. Fryer, and P. A. Berge, Measurement of transverse isotropy in marine sediments: what data are required? (abstract), *Eos, Trans. Am. Geophys. Union, 68,* 1375, 1987.

Newmark, R. L., R. N. Anderson, D. Moos, and M. D. Zoback, Sonic and ultrasonic logging of Hole 504B and its implications for the structure, porosity, and stress regimes of the upper 1 km of the oceanic crust, *Initial Rep. Deep Sea Drill. Project, 83,* 479–510, 1985.

Purdy, G. M., The variability in seismic structure of layer 2 near the East Pacific Rise at 12°N, *J. Geophys. Res., 87,* 8403–8416, 1982.

Purdy, G. M., The seismic structure of 140 Myr old crust in the western central Atlantic Ocean, *Geophys. J. R. Astron. Soc., 72,* 115–137, 1983.

Purdy, G. M., New observations of the shallow seismic structure of young oceanic crust, *J. Geophys. Res., 92,* 9351–9362, 1987.

Salisbury, M. H., and N. I. Christensen, The seismic velocity structure of a traverse through the Bay of Islands ophiolite complex, Newfoundland, an exposure of oceanic crust and upper mantle, *J. Geophys. Res., 83,* 805–817, 1978.

Schreiber, E., and P. J. Fox, Compressional wave velocities and mineralogy of fresh basalts from the Famous area and the Oceanographer Fracture Zone and the texture of layer 2A of the oceanic crust, *J. Geophys. Res., 81,* 4071–4076, 1976.

Schreiber, E., and P. J. Fox, Density and *P*-wave velocity of rocks from the FAMOUS region and their implication to the structure of the oceanic crust, *Geol. Soc. Am. Bull., 88,* 600–608, 1977.

Schoenberg, M., Reflection of elastic waves from periodically stratified media with interfacial slip, *Geophys. Prospecting, 31,* 265–292, 1983.

Shearer, P. M., Cracked media, Poisson's ratio and the structure of the upper oceanic crust, *Geophys. J., 92,* 357–362, 1988.

Shearer, P. M., and J. A. Orcutt, Anisotropy in the oceanic lithosphere – theory and observations from the Ngendei seismic refraction experiment in the south-west Pacific, *Geophys. J. R. Astron. Soc., 80,* 493–526, 1985.

Shearer, P. M., and J. A. Orcutt, Compressional and shear wave anisotropy in the oceanic lithosphere — the Ngendei seismic refraction experiment, *Geophys. J. R. Astron. Soc., 87,* 967–1003, 1986.

Spudich, P., and J. Orcutt, Petrology and porosity of an oceanic crustal site: Results from waveform modeling of seismic refraction data, *J. Geophys. Res., 85,* 1409–1433, 1980a.

Spudich, P., and J. Orcutt, A new look at the seismic velocity structure of the oceanic crust, *Rev. Geophys. and Space Phys., 18,* 627–645, 1980b.

Stephen, R. A., Seismic anisotropy observed in upper oceanic crust, *Geophys. Res. Letters, 8,* 865–868, 1981.

Stephen, R. A., Seismic anisotropy in the upper oceanic crust, *J. Geophys. Res., 90,* 11383–11396, 1985.

Stephen, R. A., Lateral heterogeneity in the upper oceanic crust at DSDP Site 504, *J. Geophys. Res., 93,* 6571–6584, 1988.

White, R. S., and R. B. Whitmarsh, An investigation of seismic anisotropy due to cracks in the upper oceanic crust at 45°N, Mid-Atlantic Ridge, *Geophys. J. R. Astron. Soc., 79,* 439–467, 1984.

Whitmarsh, R. B., Seismic refraction studies of the upper igneous crust in the North Atlantic and porosity estimates for layer 2, *Earth and Planet. Sci. Letters, 37,* 451–464, 1978.

RECENT PACIFIC-EASTER-NAZCA PLATE MOTIONS

David F. Naar[1] and R. N. Hey

Hawaii Institute of Geophysics
School of Ocean and Earth Science and Technology
University of Hawaii, Honolulu, Hawaii 96822

Abstract. Instantaneous relative plate motions have been calculated for the Pacific, Easter and Nazca plates by inverting spreading rates since the Brunhes/Matuyama reversal boundary (obtained from modeling 39 magnetic anomaly profiles across the divergent boundaries of all three plates), along with 10 transform azimuths (obtained from recent SeaMARC II, GLORIA and Sea Beam data) and 20 published seismic slip vectors. The rates along the Nazca–Pacific and Nazca–Easter spreading axes increase to the south. The rates along the Pacific–Easter spreading axis decrease to the south. Along ~2400 km of the southern Nazca–Pacific plate boundary where spreading rates range from 145 to 160 km there are no Nazca–Pacific transform faults where spreading axes are offset. Instead, the offsets are accommodated by microplates, propagating rifts, or overlapping spreading centers. The origin of the Easter microplate cannot be attributed solely to fast spreading rates along the preexisting Nazca–Pacific boundary because the fastest seafloor spreading is to the south of the microplate. The Nazca–Pacific Euler vector (0–0.73 Ma) from this study has a slower angular velocity and lies outside the confidence ellipse of the Minster and Jordan RM2 Euler vector (0–3.0 Ma). It also lies outside of the confidence ellipse of the DeMets et al. NUVEL-1 Euler vector (0–3.0 Ma) but has approximately the same angular velocity. Our preferred Euler vector describing the absolute motion of the Easter microplate is near the center of the microplate with an angular velocity of about 15°/m.y., making it a fast 'spinning' plate. Oblique convergence is predicted along the proposed Nazca–Easter and Pacific–Easter transform segments of the proposed northern and southern triple junctions, respectively. The similarity between the best fitting Euler vector for all three plate pairs and the Euler vectors derived by the three-plate closure condition suggests the microplate interior is behaving mostly rigidly. Reduced chi-squared values and F-ratio tests support this finding. However, comparison of the predicted motion vectors with the observed structures interpreted to be microplate boundaries indicates that deformation must be occurring over a broad area along the northern microplate boundary. This deformation is suspected to be a direct consequence of the large-scale rift propagation and rapid microplate rotation.

---

[1]Also at Scripps Institution of Oceanography and now at the University of South Florida, Department of Marine Science, St. Petersburg, Florida 33701

Introduction

The Easter microplate exists along the fastest divergent boundary of the Earth (Fig. 1). It is near the center of a large shallow area characterized by anomalously low surface and shear wave velocities [Woodhouse and Dziewonski, 1984], which together with the pattern of helium isotopes and light rare earth enrichments [Craig et al., 1984; Schilling et al., 1985] indicates very intense mantle convection. Hey et al. [1985] suggest this correlation is unlikely to be coincidental, and that the microplate is probably a transient feature between propagating and failing spreading axes (offset by about 300 km) and may be a modern analog for other large offset spreading axis jumps.

The geometric configuration of this area (Fig. 1) is different from the geometries observed elsewhere for small offset rift jumps as previously noted by Handschumacher et al., [1981]. This is because the southern part of the suspected failing West Rift is propagating southeastward towards the southern part of the northward propagating East Rift. This is different than what is observed at small offset propagating rift systems [e.g., Hey et al., 1986] where failing rifts are situated progressively farther away from the new rift (as one moves in the direction opposite to the propagation direction). The general geometry of the microplate is more like an enlarged overlapping spreading center (OSC) [Macdonald and Fox, 1983; Lonsdale, 1983] except that the microplate rift tips appear to be connected to transform zones.

The microplate was initially suspected to exist because of the ring-like pattern of earthquake epicenters and unusual fault plane mechanisms [Herron, 1972a; Forsyth, 1972]. Engeln and Stein [1984] were the first to investigate the relative plate motions in this area using the magnetic anomaly interpretation of Handschumacher et al. [1981] and slip vectors from the fault plane solutions derived from Forsyth [1972], Anderson et al. [1974] and their own study. Their model was based on 34 data and there were no transform azimuths and only one spreading rate for the Pacific-Easter boundary. Since then, more magnetic, bathymetric and side-scan data have become available. All the new and old magnetic anomaly data have been analyzed to obtain spreading rates during the Brunhes (Table 1). There are twice as many data than before, and there are four transform azimuths and seven spreading rates for the Pacific-Easter boundary. These new data (Table 2) have been inverted to determine the recent (0–0.73 Ma) relative motions for the three plates.

Data

A. Seismic Slip Vectors

Twenty seismic slip vectors were selected from various sources (Table 2). For the motion inversion, the assigned uncertainty for each slip vector azimuth is 20 degrees, except for those given a lower uncertainty

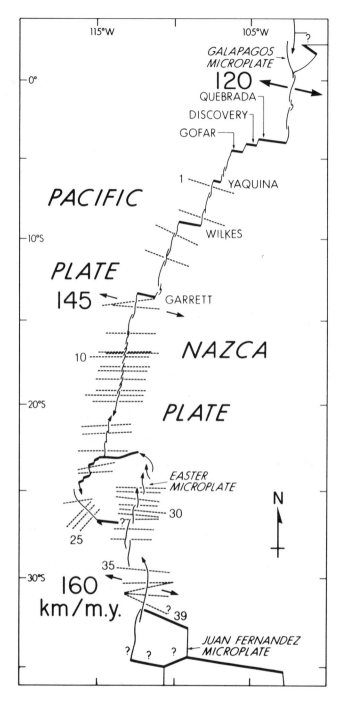

Fig. 1. Location of Easter microplate and profile index of magnetic anomaly data. Dashed lines represent approximate locations and azimuths of data profiles. Thick lines represent transform faults. Thin lines represent spreading axes. Small solid arrows arrows indicate rift tips that appear to be currently propagating. Large arrows with bold numbers are direction and magnitude of Pacific–Nazca relative motion in km/m.y. Spreading rates are from the 3P relative motion model (0–0.73 Ma). Cocos and Antarctica plates are north and south, respectively, of Nazca plate. Modified from Naar and Hey [1989].

by Minster and Jordan [1978]. Four slip vectors from Engeln and Stein [1984] derived from strike-slip fault-plane mechanisms located near transform plate boundary structures are used, but the remaining eight slip vectors from their study are not used because they do not have a strike-slip mechanism, or are not near a transform fault boundary. This was done because of the possibility that local stress directions in tectonically complex areas can be different than plate motion direction, e.g., some earthquakes may occur along faults reoriented at a ridge-transform intersection or along secondary faults within a transform zone [Francheteau et al., 1987]. Eleven selected seismic slip vectors derived by the automated centroid moment tensor method [Dziewonski et al., 1981] are used and are matched reasonably well by the predicted azimuths within the twenty degrees of uncertainty (Table 2). The remaining five slip vectors are taken from the Minster and Jordan [1978] RM2 (relative motion model number 2) data set.

B. Bathymetry

Ten simple linear fault orientations measured from SeaMARC II side-scan and bathymetry (recently collected in October and November 1987 upon the R/V *Moana Wave* on cruises MW8710 and MW8711), GLORIA side-scan and Sea Beam multi-narrow-beam echo-sounder bathymetry are used as transform fault azimuths. Uncertainty in transform azimuths (Table 2) from our SeaMARC II data (MW8711) for the Pacific–Easter plate boundary is calculated by taking the square root of the sum of the squared navigation uncertainty plus the squared data quality uncertainty, and rounding that value to the nearest integer. If GPS (Navstar satellite Global Positioning System) navigation fixes are available across a transform, a navigation uncertainty of two degrees is assigned to the azimuth datum. If not, the navigation uncertainty is calculated by rotating the trackline to the maximum assigned error limit of the two nearest surrounding satellite fixes, which gives uncertainties that vary from two to five degrees. Data quality is a subjective uncertainty which depends on the linearity and length of the transform feature and the morphology of the surrounding area. Data quality uncertainties range from three to nine degrees.

Thus, after calculations, transform azimuth uncertainties ranged from four to ten degrees (Table 2). Even though ten degrees may appear high for a side-scan azimuth uncertainty, the neighboring complex plate fabric and the possibility that the Pacific–Easter Euler vector of relative motion may have migrated during the Brunhes require us to be conservative. An azimuth uncertainty of 5 degrees is assigned to Macdonald's unpublished SeaMARC II azimuths because only azimuths measured from the longest, straightest, and simplest transform traces are used [Macdonald, pers. comm., 1988]. A four degree uncertainty is assigned to the published GLORIA transform azimuths, because of the large GLORIA mosaic covering the Quebrada, Discovery and Gofar transform systems. An eight degree uncertainty is arbitrarily assigned for the transform azimuth obtained from Sea Beam data [Lonsdale, 1983] because presumably it should be more certain than slip vector azimuths and less certain than reliable side-scan azimuths.

C. Magnetics

All tracks with decipherable magnetic anomalies crossing the spreading axes (Fig. 1) are projected onto azimuths perpendicular to the local spreading axis orientation (Table 1). The 39 magnetic anomaly profiles and synthetic models are shown in Figs. 2–6. Spreading rate uncertainty is calculated by dividing the square root of the sum of the squared navigation, picking, and data quality uncertainties, by the 0.73 Ma age of the Brunhes/Matuyama reversal boundary [Kent and Gradstein, 1986] and rounding that value to the nearest integer. The data quality uncertainty is used as a weighting factor to emphasize a good fit of the model

TABLE 1. Magnetic Model Parameters

ID	Cruise	Lat	Lon	STR	YR	EFF	BRUNHES 0.73 Ma VEL	BRUNHES 0.73 Ma ASY	JARAMILLO 0.91 Ma VEL	JARAMILLO 0.91 Ma ASY	JARAMILLO 0.98 Ma VEL	JARAMILLO 0.98 Ma ASY
01	SCAN10AR	-6.40	-107.50	014	70	0.03	142	8	156	-10	156	-10
02	POL7304	-8.30	-107.95	014	73	0.03	132	13	138	19	138	19
03	POL7304	-9.90	-110.10	018	73	0.03	142	17	160	-26	160	0
04	POL7304	-11.40	-110.50	018	73	0.02	150	-1	168	-30	180	0
05	DSDP34GC	-13.80	-112.45	014	73	0.02	146	-6	178	8	178	8
06	72110803	-14.00	-112.50	014	73	0.02	150	-2	166	-11	166	-11
07	POL7302	-15.98	-112.97	014	73	0.02	150	-2	154	0	154	0
08	V1905	-17.00	-113.12	014	63	0.02	144	5	156	-35	180	17
09	C1306	-17.01	-113.12	014	70	0.02	146	9	146	-18	160	-14
10	C1306	-17.37	-113.23	014	70	0.02	162	-10	162	7	162	7
11	C1306	-18.00	-113.30	014	70	0.02	156	-21	152	25	140	-21
12	C1306	-18.40	-113.40	014	70	0.02	148	-16	156	-16	156	-16
13	72110806	-18.85	-113.48	014	73	0.02	150	-11	170	0	170	0
14	POL7302	-19.20	-113.56	014	73	0.02	150	-11	156	-11	156	-11
15	POL7302	-19.60	-113.65	014	73	0.02	152	-7	176	-19	176	10
16	POL7302	-20.00	-113.70	014	73	0.02	154	-9	180	-13	180	-13
17	POL7302	-21.03	-114.20	014	73	0.02	162	-17	142	24	158	-10
18	POL7302	-21.35	-114.25	014	73	0.02	152	-8	164	-7	160	-7
19	PASC04WT	-22.60	-114.48	003	81	0.02	132	5	148	5	148	5
20	PASC04WT	-23.80	-115.45	355	83	0.02	118	5	122	5	122	5
21	72110807	-24.00	-115.44	355	73	0.02	112	5	118	15	118	15
22	PASC04WT	-25.90	-115.87	320	83	0.02	78	8	60	28	60	28
23	PASC04WT	-26.00	-115.80	320	83	0.02	66	4	44	-5	44	-5
24	PASC04WT	-26.29	-115.52	320	83	0.02	56	-18	52	18	52	18
25	PASC04WT	-26.40	-115.23	320	83	0.02	46	15	36	0	36	0
26	74010903	-25.00	-112.40	005	74	0.02	76	20	90	0	90	0
27	POL7303	-25.30	-112.35	005	73	0.02	84	22	120	22	120	22
28	PASC04WT	-25.72	-112.43	005	83	0.02	88	-4	124	-15	124	-15
29	74010903	-26.00	-112.50	005	74	0.02	100	-2	100	-30	100	-30
30	PASC04WT	-26.40	-112.60	005	83	0.02	116	-12	128	-26	128	-26
31	C1502	-26.50	-112.61	005	71	0.02	118	-2	132	-25	128	20
32	EN113DIG	-26.80	-112.63	005	84	0.02	126	11	166	-10	166	-10
33	74010903	-27.20	-112.75	010	74	0.02	162	11	198	11	198	11
34	ELT29	-28.30	-112.90	010	67	0.02	162	-8	136	40	136	40
35	PASC03WT	-29.90	-111.70	012	83	0.02	158	8	164	19	180	-19
36	POL7303	-30.47	-111.65	012	73	0.02	160	0	156	-20	156	-20
37	PASC03WT	-30.70	-111.85	012	83	0.02	162	13	180	7	180	7
38	YAQ7304	-31.01	-111.73	012	73	0.02	160	-9	210	-58	160	0
39	PASC03WT	-31.40	-112.00	012	83	0.02	168	2	156	1	156	1

Additional parameters for more complicated profiles:

ID	ANOM 2 1.66 Ma VEL	ANOM 2 1.66 Ma ASY	ANOM 2 1.88 Ma VEL	ANOM 2 1.88 Ma ASY	ANOM 2A 2.47 Ma VEL	ANOM 2A 2.47 Ma ASY	ANOM 2A 3.40 Ma VEL	ANOM 2A 3.40 Ma ASY	JUMP km	JUMP Ma
20	90	15	90	15						
21	110	0	110	0						
22	60	28	60	24						
23	52	0	48	6						
24	48	0	48	6						
26	72	-27	72	0	100	14	92	18	40	0.12
27	88	-26	72	0	132	20	120	30	36	0.40
28	96	-26	72	0					26	0.84
29	114	0	72	0						
30	128	-26	72	0						

ID	Profile number	VEL	Full-rate spreading velocity
STR	Strike of spreading axis	ASY	Spreading asymmetry (percent added to the west from the east)
YR	Year data was collected	JUMP	Distance of rift jump at the time of the rift jump
EFF	Effective susceptibility		

11

to the observed magnetic anomaly. Each profile is evaluated as being excellent, very good, fair or poor. The excellent profiles (textbook examples) are assigned a quality uncertainty of zero km. The very good profiles (those with a slightly complicated magnetic pattern) are assigned 2 km. The fair profiles (those with a complex magnetic and bathymetric pattern) are assigned 5 km. The poor profiles (those that are possibly misfit) have an uncertainty estimated by measuring the distance to the nearest alternative fit which, in this area of medium to fast spreading, varied from 7 to 20 km. The following profiles (Fig. 2–6) are consi-

dered excellent: 8, 14–16, 18, 20–23, 30, and 31. These are considered very good: 5, 9, 12, 13, 24, 29, 32, 35, and 36. These are considered fair: 3, 4, 6, 7, and 10. The remainder of the profiles are considered poor: 1, 2, 11, 17, 19, 25–28, 33, 34, and 37–39. The picking uncertainty (the resolution of the modeling program) for all profiles is 2 km, and the spreading direction uncertainty (converted from degrees to km) is set at 0 km for profiles near transforms and 1 km for profiles far from transforms or near the pole of relative rotation. The resulting uncertainties vary from 3 to 28 km/m.y. (Table 2).

TABLE 2.   Three Plate Closure Model and Data

	LAT.N	LON.E	DATUM	S.D.	MODEL	RES.	IMP.	REFERENCE
					NAZC PACF			
RA	−6.40	−107.50	142.0	28.0	133.3	08.7	0.014	01-SCAN10AR
	−8.30	−107.95	132.0	22.0	136.4	−04.4	0.018	02-POL7304
	−9.90	−110.00	142.0	08.0	139.4	02.6	0.098	03- "
	−11.40	−110.50	150.0	08.0	141.7	08.3	0.076	04- "
	−13.80	−112.45	146.0	04.0	145.5	00.6	0.180	05-DSDP34GC
	−14.00	−112.50	150.0	08.0	145.7	04.3	0.043	06-72110803
	−15.98	−112.97	150.0	08.0	148.2	01.8	0.028	07-POL7302
	−17.00	−113.12	144.0	07.0	149.4	−05.4	0.030	08-V1905
	−17.01	−113.12	146.0	04.0	149.4	−03.4	0.093	09-C1306
	−17.37	−113.23	162.0	08.0	149.8	12.2	0.022	10- "
	−18.00	−113.30	156.0	25.0	150.5	05.5	0.002	11- "
	−18.40	−113.40	148.0	04.0	150.9	−02.9	0.073	12- "
	−18.85	−113.48	150.0	04.0	151.4	−01.4	0.069	13-72110806
	−19.20	−113.56	150.0	03.0	151.8	−01.8	0.119	14-POL7302
	−19.60	−113.65	152.0	03.0	152.2	−00.2	0.116	15- "
	−20.00	−113.70	154.0	03.0	152.6	01.4	0.114	16- "
	−21.03	−114.20	162.0	14.0	153.6	08.4	0.005	17- "
	−21.35	−114.25	152.0	03.0	153.9	−01.9	0.123	18- "
	−27.20	−112.75	162.0	19.0	158.1	03.9	0.008	33-74010903
	−28.30	−112.90	162.0	26.0	158.9	03.1	0.005	34-ELT29
	−29.90	−111.70	158.0	04.0	159.6	−01.6	0.285	35-PASC03WT
	−30.47	−111.65	160.0	04.0	159.9	00.1	0.312	36-POL7303
	−30.70	−111.85	162.0	18.0	160.0	02.0	0.016	37-PASC03WT
	−31.01	−111.73	160.0	26.0	160.2	−00.2	0.008	38-YAQ7304
	−31.40	−112.00	168.0	10.0	160.4	07.6	0.059	39-PASC03WT
TF	−3.70	−103.20	S78E	4.0	S80E	−2.0	0.201	Searle (1983)
	−4.00	−104.00	S78E	4.0	S79E	−1.4	0.198	"
	−4.60	−105.50	S78E	4.0	S78E	−0.2	0.195	"
	−6.25	−107.25	S76E	8.0	S77E	−1.1	0.047	Lonsdale (1983)
	−9.00	−109.30	S78E	5.0	S76E	2.1	0.115	SEAMARC II (MW8710) !
	−13.50	−111.50	S74E	5.0	S75E	−0.9	0.107	" " " !
SV	−4.40	−105.90	S75E	20.0	S78E	−2.9	0.008	Anderson et al. (1974)
	−4.50	−106.00	S76E	15.0	S78E	−1.8	0.014	Anderson & Sclater (1972)
	−4.60	−105.80	S77E	15.0	S78E	−1.0	0.014	"
	−13.30	−111.50	S75E	20.0	S75E	0.1	0.007	"
	−28.70	−112.70	S62E	20.0	S75E	−13.5	0.005	Anderson et al. (1974)
	−4.41	−105.85	S82E	20.0	S78E	4.1	0.008	Dziewonski et al. (1984A #92)
	−4.54	−104.71	S86E	20.0	S79E	7.1	0.008	" " " (1987A#123)
	−4.56	−105.41	S84E	20.0	S78E	5.7	0.008	" " " (1985 #15)
	−4.93	−105.22	S86E	20.0	S78E	7.5	0.008	" " " (1987C#371)
	−9.15	−108.44	S66E	20.0	S77E	−10.6	0.007	" " " (1987D#176)
	−12.97	−112.69	S78E	20.0	S74E	4.0	0.007	" " " (1987D#284)
	−13.33	−111.54	S82E	20.0	S75E	7.1	0.007	" " " (1987C#234)
	−13.52	−111.32	S82E	20.0	S75E	6.9	0.007	" " " (1987A #10)
	−29.07	−112.75	S75E	20.0	S75E	−0.5	0.005	" " " (1984B #64)
	−29.28	−112.71	S83E	20.0	S76E	7.5	0.005	" " " (1983 #284)
	−29.52	−112.45	S72E	20.0	S76E	−3.7	0.005	" " " (1987B#224)

TABLE 2.   (continued)

	LAT.N	LON.E	DATUM	S.D.	MODEL	RES.	IMP.	REFERENCE

EAST PACF

	LAT.N	LON.E	DATUM	S.D.	MODEL	RES.	IMP.	REFERENCE
RA	−22.60	−114.48	132.0	28.0	146.8	−14.8	0.013	19-PASC04WT
	−23.80	−115.45	118.0	03.0	118.4	−00.4	0.454	20- "
	−24.00	−115.44	112.0	03.0	113.0	−01.0	0.381	21-72110807
	−25.90	−115.87	78.0	03.0	71.3	06.7	0.236	22-PASC04WT
	−26.00	−115.80	66.0	03.0	68.0	−02.0	0.244	23- "
	−26.29	−115.52	56.0	04.0	57.2	−01.2	0.162	24- "
	−26.40	−115.23	46.0	14.0	50.0	−04.0	0.015	25- "
TF	−23.05	−114.70	N89E	4.0	N82E	−6.7	0.092	MW8711 UNPUBL. SEAMARC II
	−23.35	−115.12	N76E	10.0	N77E	1.1	0.016	" " " "
	−24.15	−115.50	N72E	4.0	N70E	−2.4	0.141	" " " "
	−24.33	−115.72	N72E	7.0	N66E	−6.1	0.049	" " " "
SV	−26.51	−114.11	N77E	20.0	N85E	7.6	0.052	Engeln and Stein (1984 #07)
	−26.67	−113.96	S71E	20.0	S89E	−18.2	0.070	" " " (" #06)
	−26.87	−113.58	S72E	20.0	S69E	3.3	0.101	" " " (" #12)
	−27.04	−113.14	S72E	20.0	S45E	27.4	0.102	" " " (" #05)

EAST NAZC

	LAT.N	LON.E	DATUM	S.D.	MODEL	RES.	IMP.	REFERENCE
RA	−25.00	−112.40	76.0	10.0	72.9	03.1	0.027	26-74010903
	−25.30	−112.35	84.0	13.0	81.7	02.3	0.013	27-POL7303
	−25.72	−112.43	88.0	11.0	93.8	−05.8	0.017	28-PASC04WT
	−26.00	−112.50	100.0	04.0	102.0	−02.0	0.133	29-74010903
	−26.40	−112.60	116.0	03.0	113.7	02.3	0.280	30-PASC04WT
	−26.50	−112.61	118.0	03.0	116.6	01.4	0.297	31-C1502
	−26.80	−112.63	126.0	04.0	125.3	00.7	0.205	32-EN113DIG +

S.D.	Datum uncertainty (one standard deviation)	TF	Transform azimuth in degrees
RES.	Residual between datum and model	SV	Seismic slip vector in degrees
IMP.	Datum importance to model (total sums to 6)	!	Unpublished data from Macdonald (UCSB)
RA	Full spreading rate in km/Ma for 0–0.73 Ma	+	Unpublished data from Schilling (URI)

Magnetic Modeling

A. Methods

The magnetic anomaly data (discussed in previous section) were modeled by generating a synthetic magnetic anomaly using the observed bathymetry, the orientation of magnetic stripes, and the magnetic polarity time scale of Kent and Gradstein [1986] modified to include the Emperor reversed event [Ryan, 1972; Champion et al., 1981] which Wilson and Hey [1981] date at 0.49–0.50 Ma based on modeling magnetic anomalies along the Cocos–Nazca spreading axis. All the parameters used for the models are listed in Table 1. To focus on the young anomalies, the profiles in Figs. 2–6 do not extend beyond 200 km. Ideally, using rates calculated from modeling the younger Emperor reversed event would give a more accurate estimate of the more recent plate motions. Unfortunately, it is not observed in all the profiles, especially in areas of slower spreading. The rate data from RM2 or NUVEL-1 are not used because they are averaged over a longer time interval (0–3 Ma). Thus, only the spreading rates derived from modeling the Brunhes/Matuyama reversal boundary are inverted for the relative and absolute rotation Euler vectors.

The location of the spreading axis is defined by bathymetry and, if observed, by the high-amplitude axial magnetic anomaly which appears to exist along most spreading axes [Rea and Blakely, 1975; Wilson and Hey, 1981]. The skewness of some axial magnetic anomalies (e.g., Fig. 4: profiles 22–24) is an additional check on the spreading direction used to generate the magnetic model. We do not model recent rift jumps within areas of seafloor formed during the Brunhes. Instead, asymmetry is used to match what could either be asymmetric spreading or asymmetric accretion (local rift jumping most likely resulting from rift propagation). The rates used to generate the magnetic models beyond the age of the Brunhes/Matuyama reversal boundary (Figs. 2–6) are listed in Table 1.

Identification of the magnetic anomalies along profiles across the northeast divergent boundary of the microplate (north of 25 °S) is difficult because of extremely rugged bathymetry. Thus, four tenuous profiles from this area are not included even though the predicted motion vectors approximately match the rates measured in two of the four profiles. Near 25 °S, 112 °25 ′W, there is a propagating rift tip that has been propagating northward throughout the duration of the Brunhes into seafloor formed before the Brunhes [Naar and Hey, 1986]. SeaMARC II data collected in this area show there is a local failed rift 45 km to the east of this propagating rift, and the magnetic models have been based on this interpretation. The total spreading rates for these profiles result from summing the rates from both the failed and propagating rifts and are given high uncertainties because of the uncertain identification of the western Jaramillo due to the extra magnetic wiggles to the west of the western pseudofault (western-most dashed line in Fig. 5: profiles 26–28).

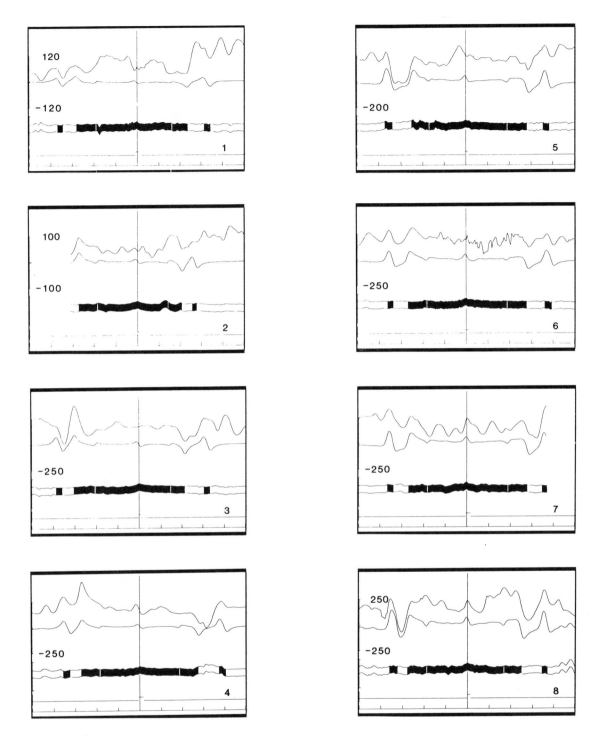

Fig. 2. Magnetic anomaly data and synthetic models for profiles indexed in Fig. 1 arranged from north to south. Full spreading rates used to match the 0.73 Ma Brunhes/Matuyama reversal are listed in Table 1 and 2. For each set, top profile is data (nT), middle profile is model (nT), lower profile is bathymetry (m), darkened blocks within 1-km magnetic layer following bathymetry are blocks of normal magnetic polarity, horizontal line represents magnetic intensity used (see Table 1) except double intensity spike for 0–0.02 Ma at ridge axis, lower scale is distance (from -100 to 100 km, each division is 20 km) and vertical scale bars are for magnetics and bathymetry (2–6 km). Each model is generated using bathymetry, ridge orientation, polarity, and magnetic intensity and additional parameters listed in Table 1.

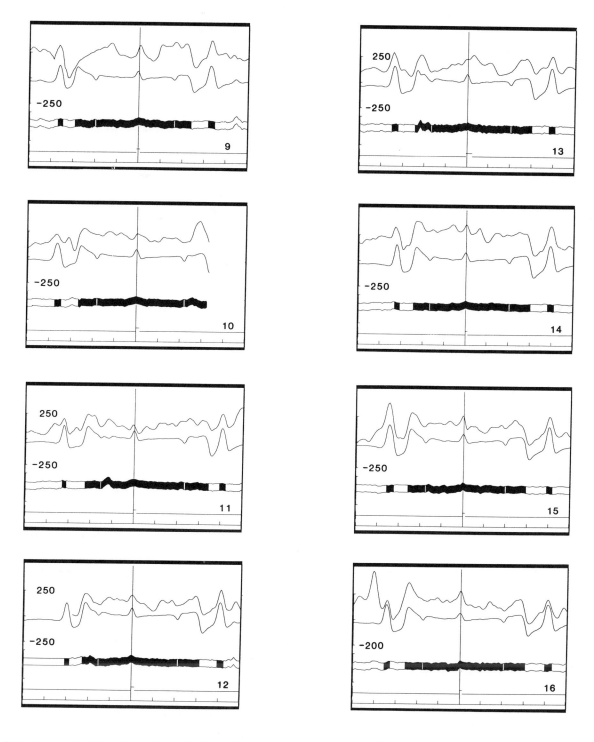

Fig. 3. Magnetic anomaly data and synthetic models for profiles indexed in Fig. 1 arranged from north to south. Full spreading rates used to match the 0.73 Ma Brunhes/Matuyama reversal are listed in Table 1 and 2. For each set, top profile is data (nT), middle profile is model (nT), lower profile is bathymetry (m), darkened blocks within 1-km magnetic layer following bathymetry are blocks of normal magnetic polarity, horizontal line represents magnetic intensity used (see Table 1) except double intensity spike for 0–0.02 Ma at ridge axis, lower scale is distance (from -100 to 100 km, each division is 20 km) and vertical scale bars are for magnetics and bathymetry (2–6 km). Each model is generated using bathymetry, ridge orientation, polarity, and magnetic intensity and additional parameters listed in Table 1.

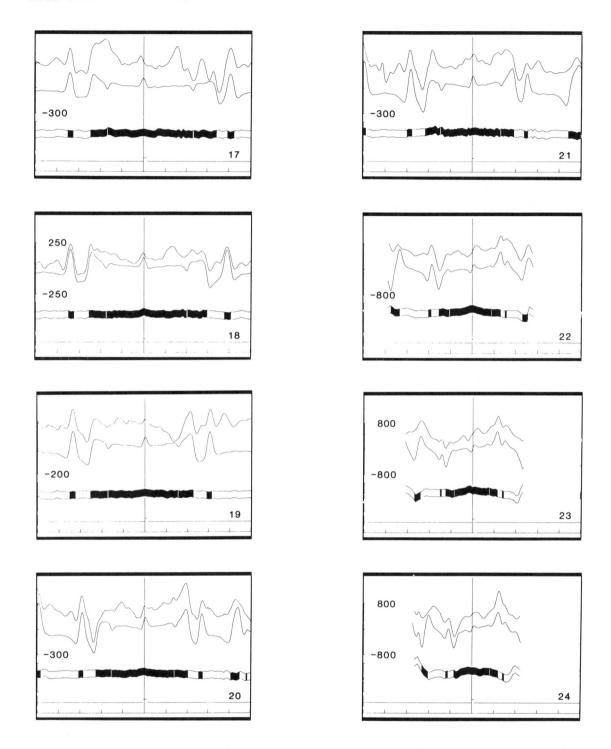

Fig. 4. Magnetic anomaly data and synthetic models for profiles indexed in Fig. 1 arranged from north to south. Full spreading rates used to match the 0.73 Ma Brunhes/Matuyama reversal are listed in Table 1 and 2. For each set, top profile is data (nT), middle profile is model (nT), lower profile is bathymetry (m), darkened blocks within 1-km magnetic layer following bathymetry are blocks of normal magnetic polarity, horizontal line represents magnetic intensity used (see Table 1) except double intensity spike for 0–0.02 Ma at ridge axis, lower scale is distance (from -100 to 100 km, each division is 20 km) and vertical scale bars are for magnetics and bathymetry (2–6 km). Each model is generated using bathymetry, ridge orientation, polarity, and magnetic intensity and additional parameters listed in Table 1.

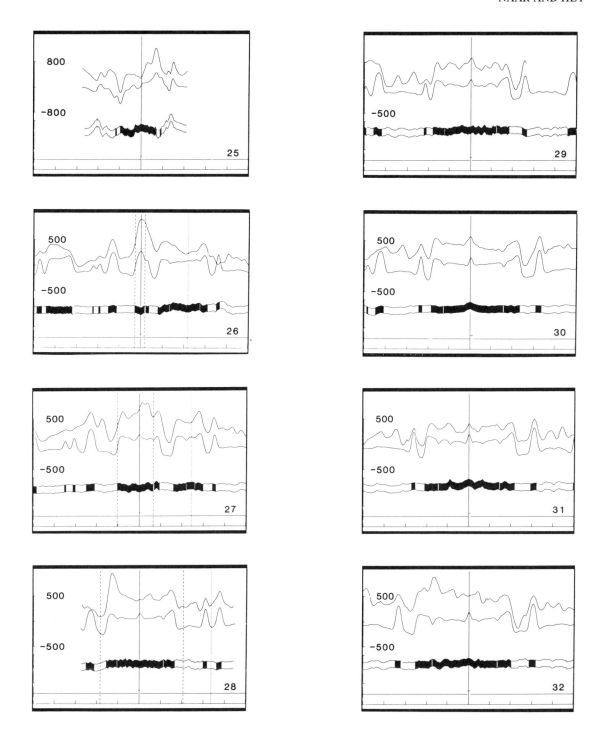

Fig. 5. Magnetic anomaly data and synthetic models for profiles indexed in Fig. 1 arranged from north to south. Full spreading rates used to match the 0.73 Ma Brunhes/Matuyama reversal are listed in Table 1 and 2. For each set, top profile is data (nT), middle profile is model (nT), lower profile is bathymetry (m), darkened blocks within 1-km magnetic layer following bathymetry are blocks of normal magnetic polarity, horizontal line represents magnetic intensity used (see Table 1) except double intensity spike for 0–0.02 Ma at ridge axis, lower scale is distance (from -100 to 100 km, each division is 20 km) and vertical scale bars are for magnetics and bathymetry (2–6 km). Each model is generated using bathymetry, ridge orientation, polarity, and magnetic intensity and additional parameters listed in Table 1.

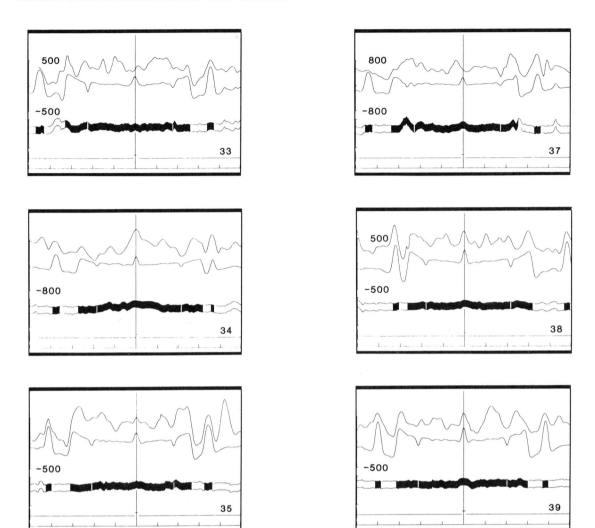

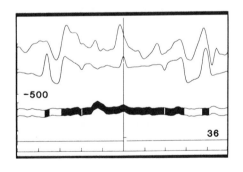

Fig. 6. Magnetic anomaly data and synthetic models for profiles indexed in Fig. 1 arranged from north to south. Full spreading rates used to match the 0.73 Ma Brunhes/Matuyama reversal are listed in Table 1 and 2. For each set, top profile is data (nT), middle profile is model (nT), lower profile is bathymetry (m), darkened blocks within 1-km magnetic layer following bathymetry are blocks of normal magnetic polarity, horizontal line represents magnetic intensity used (see Table 1) except double intensity spike for 0–0.02 Ma at ridge axis, lower scale is distance (from -100 to 100 km, each division is 20 km) and vertical scale bars are for magnetics and bathymetry (2–6 km). Each model is generated using bathymetry, ridge orientation, polarity, and magnetic intensity and additional parameters listed in Table 1.

B. Assessment of Rates Obtained from Magnetic Modeling

The rate data (Fig. 7) delineate the divergent boundaries of the three plates and are consistent with each other (except for a few cases). The rates required to match the Brunhes/Matuyama reversal do not vary as much as the rates required to match the Jaramillo from one profile to another (Table 1). In favor of brevity, detailed assessment of each profile is not made except for the evaluation of data quality in the previous data section. In general, the rates are in agreement with recent Nazca–Pacific spreading rate estimates [DeMets et al., 1989; Gordon et al., 1988; Macdonald et al., 1988].

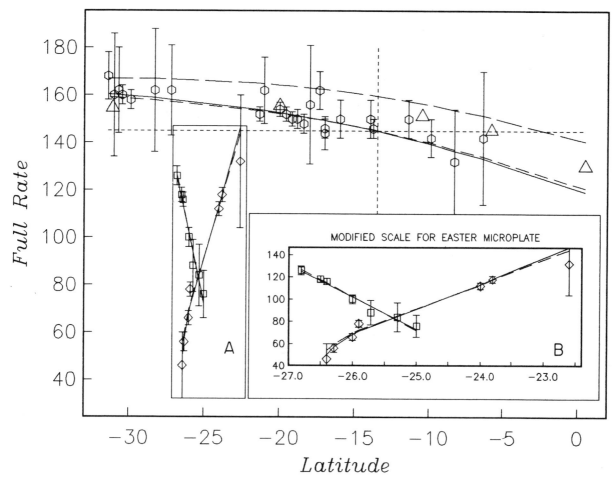

Fig. 7. Full spreading rate versus latitude. Rates are in km/m.y. Latitude is in degrees. Hexagons represent Nazca–Pacific rates during the Brunhes obtained and used in this study. Diamonds represent Pacific–Easter rates. Squares represent Nazca–Easter rates. Error bars represent assigned standard errors (see text). Large triangles represent averaged rates normalized to 0.73 Ma (Brunhes) from regional studies [Rea and Scheidegger, 1979; Lonsdale, 1988]. These five data values are shown for reference but are not used. Solid curves represent 3P motion model. Long-dashed curve represents normalized RM2 rates. Dashed curves represent best fitting Euler vector from this study. Short-dashed vertical and horizontal lines represent the location of the Garrett transform fault and 145 km/m.y. There are no transform faults that exist with slip rates greater than this value along the EPR. Box B has a stretched latitude scale and a compressed rate scale to highlight the fit of models to data shown in box A.

C. Speed Limit for a Spreading Axis

The new rates are also similar to published rate values from Rea and Scheidegger [1979], and Lonsdale [1988] (large triangles in Fig. 7) and are slower than the predicted rates from RM2 (long-dashed line in Fig. 7) when all these rates are normalized to the Kent and Gradstein [1986] time scale. The rates are normalized by multiplying them by the age of the Brunhes/Matuyama reversal previously used and dividing them by the 0.73 Ma age used in this study. The RM2 predicted rates were normalized by multiplying them by the 3.32 Ma age of anomaly 2′ [Talwani et al., 1971] and dividing them by the 3.40 age of anomaly 2′ [Kent and Gradstein, 1986]. The new rates indicate the 'equator' of fast Nazca–Pacific spreading is farther south of the microplate than the 'equator' of the RM2 Euler vector for Nazca–Pacific motion (Fig. 7). Thus, the speculative hypothesis, discussed in Hey et al. [1985], that the Easter microplate formed because the separation rate of the Pacific and

Nazca plates was too fast for only one spreading axis to generate enough lithosphere is shown not to be true. However, this does not rule out the possibility that hot, intense mantle convection may thin and weaken the lithosphere to such an extent that multiple spreading axes are required when spreading rates become greater than some limiting value [Hey et al., 1985].

D. Speed Limit for the Existence of a Transform Fault

The new rates, new SeaMARC II data (MW8711) south of the microplate and previously collected Sea Beam data along the East Pacific Rise (EPR) [Figure 1 of Macdonald et al., 1986] show that transform faults do not currently exist at spreading rates greater than about 145 km/m.y. (horizontal short dashed line in Fig. 7). Herron [1972a, b] and subsequent authors have proposed one or more transform faults to exist south of the Easter microplate, near 29°S, on the basis of the earthquake epicenter distribution and offset magnetic anomalies. These transform faults were not observed in the SeaMARC II data [Hey et al., 1988; Klaus et al., 1989] or previously collected Sea Beam data [Francheteau et al., 1987]. Instead, a large overlapping rift system is observed south of the microplate near 29 °S. The lack of transform faults along the EPR between the Garrett transform fault at 13.5 °S (vertical short dashed line in Fig. 7) and the microplate is further corroborated by unpublished SeaMARC II data [MW8710, Macdonald, pers. comm., 1988]. This new finding may create a bias in the correlation of fracture zone spacing to spreading rate [Sandwell, 1986] because the large gap of fracture zones south of the Garrett Fracture Zone may be due to factors other than reduced thermal contraction. Our finding does not, however, affect the findings of Abbott [1986] which were based on spreading segment lengths from areas other than the Nazca–Pacific plate boundary.

Because of this observation and wax modeling observations [Naar et al., 1986; Oldenburg and Brune, 1972, 1975], we hypothesize that there is a speed limit of ~145 km/m.y. above which transform faults cannot exist, and instead spreading segments are offset by non-transform features (e.g., microplates, propagating rift systems, or overlapping spreading centers). Other areas of fast seafloor spreading preserved on older seafloor should be investigated to test this hypothesis, which will require high-resolution bathymetry or side-scan data in areas where magnetic isochrons are offset. For further discussion see Naar and Hey [1989].

E. The Emperor Reversed Event

The 0.49–0.50 Ma Emperor reversed event [Ryan, 1972; Champion et al., 1981; Wilson and Hey, 1981] is generally well-matched in most of the profiles in areas of fast spreading (when using the rates obtained from modeling the Brunhes). Modeling profiles with a suspected rift jump (instead of using asymmetry) within the Brunhes normal period produces a better fit to the Emperor reversed event in some of the profiles. This was not done, however, because of the difficulty in constraining a rift jump within the Brunhes and claiming that one small-wavelength magnetic anomaly is the reversed Emperor event (as opposed to small-wavelength noise) without some other kind of age information. In places of asymmetric spreading, the Emperor event was usually more visible and better matched on the faster side (as expected).

The Emperor reversed event is matched (Figs. 2–6) in the following Pacific–Easter profiles: 20 (east better than west), 21 (only on the east), 22 (on both sides), and 23 (only on the west). The Emperor event is not matched in profiles 24 and 25, probably because of slow spreading. Profile 19 has a very broad negative anomaly on the east, too wide to be attributed solely to the Emperor event, but has a good fit to the west. For the Nazca–Easter profiles, the Emperor event is matched in profiles 26 (surrounding the failed rift; west better than east), 30–32 (east better than west), but not in profiles 27–29. For the Nazca–Pacific profiles, we obtain fits (some better than others) to the Emperor event in profiles 5–7, 11–13, 15–17, 33–39. The remaining Nazca–Pacific profiles show no evidence of the Emperor reversed event.

F. Comparison with Previous Work

Naar and Hey [1986] mapped and modeled the magnetics of a fast northward propagating rift along the East Rift from 26°30 'S to 25 °S. The magnetic anomalies outside of the new wedge of lithosphere created by this fast propagating rift were not identified so they assumed a constant full spreading rate of 120 km/m.y. up to the latitude of the propagator tip in calculating the recent propagation rate of 150 km/m.y. Sempere et al. [1989] performed a three-dimensional inversion of the magnetic field of the tip of this fast propagator, obtained a ratio of propagation to full spreading rate equal to 1.5 (slightly larger than the previous value of 1.3) and calculated a revised propagation rate esti-

mate of 185 km/m.y. using the previous 120 km/m.y. spreading rate. The older anomalies surrounding this propagator are now tentatively identified and provide new spreading rates to constrain the recent propagation rate. A new estimate for the recent propagation rate near the tip is 120 km/m.y. (1.5 × ~80 km/m.y.). The 150 km/m.y. propagation rate is still valid for the long-term propagation rate during the last million years because it was calculated by dividing the distance of propagation during the Brunhes by the 0.73 Ma age of the Brunhes and thus is independent of spreading rate. The 180 km/m.y. rates published for profile 34 [Herron, 1972b; Naar and Hey, 1986] resulted from matching anomaly 3. Modeling only out to the Brunhes/Matuyama reversal gives 162 km/m.y. Some rates in this study are different from Naar and Hey [1986] because only one rate was used to match both the Brunhes and Jaramillo anomalies, whereas in this study, each reversal boundary was modeled.

Inversion for Relative Motions

A. Methods

Two kinds of inversions are made. One is to obtain the best fitting Euler vector (BFP) for each plate pair. The other is for all three plates which requires that the three Euler vectors add to zero (closure condition). We call this our 3P (three-plate) model. We used the Minster and Jordan maximum likelihood inversion algorithm. Simply stated, their algorithm repeatedly tries to minimize the squared residuals (data minus synthetic values) by adjusting the trial Euler vectors. For a complete description see Minster et al. [1974] and Minster and Jordan [1978]. When discussing relative motions, the first plate moves with respect to the second plate.

To obtain Euler vectors of relative motion, starting Euler vectors must be used, so we started with those previously published for the various plate pairs [Minster and Jordan, 1978; and Engeln and Stein, 1984]. There are no transform data for the Nazca–Easter (NAZ–EAS) boundary. The orientation of the long proposed northern transform boundary [Hey et al., 1985; Francheteau et al., 1988] was not used as transform data because of its long length and its uncertain origin. A motion direction of N87W for the southern-most datum on the NAZ–EAS boundary is used in the inversion to obtain the NAZ–EAS BFP. One azimuth was needed to make the algorithm converge for these two plates and was obtained by assuming that the relative plate motion is orthogonal to the orientation of the shallow East Rift near 26.5 °S [Plate 1 of Naar and Hey, 1986; Baker et al., 1988; Hagen et al., 1989]. This azimuth has an importance of nearly one for the BFP model, so to prevent circular reasoning, it is not used to derive the 3P model. Instead, the NAZ–EAS motion direction is calculated from the three-plate closure condition.

B. Assessment of the Relative Motion Models

The predicted rates, azimuths and slip vectors, the standard deviation of each datum, the residuals (the datum minus the model), and data importances of the 3P motion model are listed in Table 2. Six iterations were required to invert the 69 data to obtain the new relative Euler vectors for the three plate pairs. A grid search about the Euler vectors for alternative Euler vectors was done and none were found. This indicates that the Euler vectors in this study do not result from a local minimum of squared errors as discussed in Minster and Jordan [1978]. Further tests were made using various combinations of the data, e.g., rates only, rates and slip vectors, rates and transform azimuths, slip vectors only, transform azimuths only, and slip vectors and transform azimuths. The results of these tests indicate that the rate data are the most important data used to derive the new Euler vectors. This is also reflected by the importance values listed for the data in Table 2.

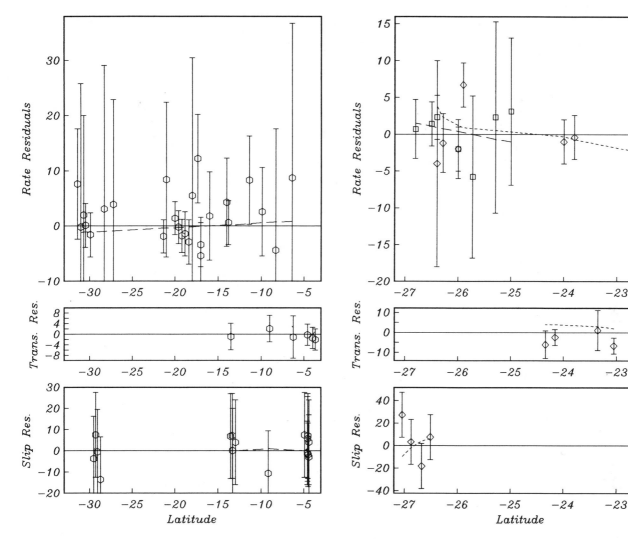

Fig. 8. Rate, transform and seismic slip residuals for Nazca–Pacific boundary versus latitude. Dashed line represent best fitting Euler vector model. Symbols described in Fig. 7. Residuals equal data minus model values and are in km/m.y. (rates) and degrees (slip and transform azimuths).

Fig. 9. Rate, transform and seismic slip residuals versus latitude for Pacific-Easter and Nazca-Easter boundaries. Short and long dashed lines represent best fitting Euler vector model for Pacific-Easter and Nazca-Easter boundaries, respectively. Symbols and units as in Figs. 7 and 8.

Spreading rate data and the 3P and BFP models are plotted with respect to latitude for all three plate pairs (Fig. 7). All residuals shown in Table 2 are also plotted with respect to latitude for the Nazca–Pacific (NAZ–PAC) plate pair (Fig. 8) and for the plate pairs associated with the Easter microplate (Fig. 9). The new NAZ–PAC Euler vectors (3P and BFP) and their associated errors along with other previously published Euler vectors are given in Table 3 and shown in Fig. 10. The 3P and RM2 NAZ–PAC relative Euler vectors are mutually exclusive but near each other (Fig. 10). The NAZ–PAC 3P Euler vector is very close to the NAZ–PAC BFP Euler vector as are the Minster and Jordan global and best fitting poles (RM2 and RBF in Fig. 10). The NAZ–EAS and PAC–EAS relative Euler vectors, the predicted motion vectors along the microplate boundaries, and the velocity triangles for the two proposed triple junctions are shown in Fig. 11 and will be discussed in the following sections. The inversion algorithm tries to simultaneously fit all the

data and thus if there is an anomalous pattern in the residuals (e.g., skewness about the zero-line) it indicates there is an inconsistency in the data set.

For the NAZ–PAC boundary (Fig. 8), there is a scatter of about 18 km/m.y. for the rate residuals, 5° for the transform residuals, and 22° for the slip vector residuals. The rate residuals are somewhat skewed so that the model may be underestimating the rates north of 15°S. For the BFP model, the scatter is about the same but the rates are slightly faster north of 15°S and slightly slower south of the microplate. The transform azimuths may be slightly underestimated in the north, yet the slip vector residuals appear evenly distributed about the zero-line. The BFP transform and seismic slip residuals are nearly the same as the 3P residuals. Matching the rates to the north of 15°S requires that the transform and seismic slip residuals increase. There are only 2 out of 47 data which are not fit within their uncertainty. Thus, there is an

TABLE 3.   Rotation Poles

		Relative Poles of Rotation				
ID	MOVING	FIXED-PLATE	LAT.	LONG.	ANG.VEL.	RCS
		This Paper				
3P	NAZC	PACF	46.77	−91.76	0.1463E+01	0.392
3P	NAZC	EAST	−22.49	−112.41	0.1499E+02	0.392
3P	PACF	EAST	−27.77	−113.98	0.1460E+02	0.392
BFP	NAZC	PACF	48.12	−90.50	0.1446E+01	0.229
BFP	NAZC	EAST	−22.63	−112.36	0.1567E+02	0.282
BFP	PACF	EAST	−28.28	−114.13	0.1311E+02	0.793
		Engeln and Stein (1984)				
ENS	NAZC	PACF	55.20	−86.70	0.1500E+01	0.550
ENS	NAZC	EAST	−20.90	−111.50	0.1160E+02	0.550
ENS	PACF	EAST	−28.30	−113.60	0.1140E+02	0.550
		Minster and Jordan (1978)				
RM2	NAZC	PACF	56.64	−87.88	0.1539E+01	0.363
BFP	NAZC	PACF	55.64	−85.76	0.1527E+01	0.334
		Demets et al. (1989)				
NUV	NAZC	PACF	55.60	−90.10	0.1420E+01	0.241
BFP	NAZC	PACF	53.80	−88.20	0.1420E+01	0.300

				Absolute Poles of Rotation			
	PLATE	LAT.	ERROR	LONG.	ERROR	OMEGA DEG./M.Y.	ERROR (1s)
				This Paper (1 Ma)			
EPP	PACF	−36.00	0.00	104.00	0.00	0.8400E+00	0.0000E+00
3P1	NAZC	55.46	8.70	−119.67	5.77	0.6951E+00	0.2758E−01
3P1	EAST	25.11	0.06	67.80	0.15	0.1486E+02	0.5846E+00
				This Paper (3.2 Ma)			
POL	PACF	−60.60	0.00	84.90	0.00	0.9850E+00	0.0000E+00
3P3	NAZC	21.75	7.50	−88.99	5.20	0.5601E+00	0.7629E−01
3P3	EAST	23.90	0.08	66.16	0.12	0.1464E+02	0.5848E+00
				This Paper (10 Ma)			
A12	PACF	−61.66	0.00	97.19	0.00	0.9670E+00	0.0000E+00
3P0	NAZC	21.25	7.25	−99.17	3.95	0.5939E+00	0.7982E−01
3P0	EAST	24.08	0.07	67.05	0.13	0.1458E+02	0.5847E+00
				Minster and Jordan (1978)			
A12	NAZC	47.99	9.36	−93.81	8.14	0.5850E+00	0.9700E−01

RCS	Reduced chi-square value (see text for description)	POL	0–3.2 Ma PACF-hotspot pole (Pollitz, 1986)
BFP	Best fitting pole for 2-plate inversion	A12	0–10 Ma PACF-hotspot pole (Minster and Jordan, 1978)
EPP	0–0.1 Ma PACF-hotspot pole (Epp, 1978)		

inconsistency in the data set, but it is small and lies within the uncertainty of most of the data.

For the NAZ-EAS boundary (Fig. 9), there is a scatter of about 8 km/m.y. for the rate residuals. There is a slight skewness of the rate residuals from the 3P model which results from the directional information from the other two plates. The BFP predicted rates are slightly faster in the south and slightly slower in the north than the 3P predicted rates.

For the PAC-EAS boundary (Fig. 9), there is a scatter of about 22 km/m.y., 8°, and 44°, for the rate, transform, and seismic slip vector residuals. In general, there is no apparent skewness for these rate residuals. The rate datum that has the highest residual (near 23 °S) is from profile 19 (Fig. 9). This profile gives a rate which is slower than the predicted rates from both the NAZ–PAC and the PAC–EAS Euler vectors (Fig. 7). This rate is included with the PAC–EAS data set instead of the NAZ–PAC data set because it is much slower than the other nearby NAZ–PAC rates derived from the excellent profiles 14–16, and 18 (Figs. 3 and 4). It is also slower than the predicted PAC–EAS rates (Fig. 7), which suggests there are local complexities in this region just to the north of the Easter microplate (large stippled area in Fig. 11). The trans-

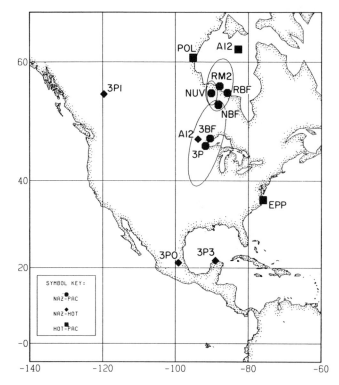

Fig. 10. Relative and absolute Euler poles for Pacific (PAC) and Nazca (NAZ) plates. 3P is our NAZ–PAC relative Euler pole derived from three plates. 3BF is our best fitting pole for NAZ–PAC boundary. RM2 and RBF are Minster and Jordan [1978] global inversion Euler pole and best fitting pole for NAZ–PAC boundary. NUV (NUVEL-1) and NBF are DeMets et al. [1989] global inversion Euler pole and best fitting pole for NAZ–PAC boundary. A12 (solid diamond) is AM1-2 absolute NAZ pole. 3P1 is absolute Euler pole for NAZ derived by adding 3P NAZ–PAC Euler vector to 0–1.0 Ma PAC absolute Euler vector [Epp, 1978] shown as EPP. 3P3 Euler pole is (preferred) NAZ absolute pole derived by adding 3P NAZ–PAC Euler vector to 0–3.2 Ma Euler vector (Pollitz, 1988) shown as POL. 3P0 pole results from adding 3P NAZ–PAC Euler vector to Minster and Jordan [1978] AM1-2 PAC absolute Euler vector (0–10.0 Ma) shown as A12 (solid square). The 95% (2-sigma) error ellipses are for 3P, NUV, and RM2 NAZ–PAC vectors.

form azimuths are slightly overestimated and the slip azimuths are slightly underestimated by the 3P model. Taken together they have equal scatter about the zero-line, except that all the transform data are in the north and all the slip vector data are in the south, so there is an inherent imbalance of the data distribution. The BFP for PAC–EAS has slower rates for the northern profiles and faster rates for the southern profiles than the 3P model. The BFP does not fit the transform and seismic slip azimuths as well as the 3P model. Thus there is some inconsistency in the data for the PAC–EAS boundary which is primarily due to discrepancies between the rate and direction data.

Mismatch between individual datum and predicted motion vectors could be due to several factors. The magnetic anomaly data are complex in some areas and may be misinterpreted, especially in the profiles with a high uncertainty (Table 2). It is also probable the Euler vectors have not been constant during the past 0.73 m.y. Although most of the seismic slip data are poorly constrained (possibly due to the remote location, the anomalous seismic velocity structure of this area, and the

small body-wave magnitudes), they are assigned large uncertainties and do not appear to be a significant source of error. Furthermore, the plate motion directions rely more on the well-constrained transform fault azimuths derived from the SeaMARC II and Sea Beam systems than the seismic slip vectors.

To statistically assess the new Euler vectors, we have applied the reduced chi-squared test and F-ratio test to both the BFP and the 3P models. This was done in the manner described by Minster and Jordan [1978] and Gordon et al. [1987]. The chi-squared value equals the sum of squared weighted residuals (i.e., residuals divided by their corresponding standard errors). The reduced chi-squared value is the ratio of chi-squared to the number of degrees of freedom, $N - 3p$, where N represents the number of data and p represents the number of independent Euler vectors (each with three components) estimated from the data [Gordon et al., 1987]. Reduced chi-squared values near 1 suggest good fits of the model to the data. Values much greater than 1 suggest poor fits. Values much less than 1 do not necessarily suggest better fits, because of the possibility of overestimating the standard errors of the data [e.g., Minster and Jordan, 1978].

For our 3P model: $N = 69$ and $p = 2$ (the third Euler vector is dependent on the other two Euler vectors because of the closure condition) which results in 63 degrees of freedom. For our NAZ–PAC BFP model: $N = 47$ and $p = 1$ which results in 44 degrees of freedom. For our PAC–EAS BFP model: $N = 15$ and $p = 1$ which results in 12 degrees of freedom. For our NAZ–EAS BFP model: $N = 8$ (as previously discussed, one orthogonal transform azimuth was added to the 7 rate data when the BFP was estimated) and $p = 1$ which results in 5 degrees of freedom. The reduced chi-squared value for our 3P model (for all three relative motion Euler vectors) is 0.392. The reduced chi-squared values for the NAZ–PAC, PAC–EAS, NAZ–EAS BFP models are 0.229, 0.793, and 0.282. These low values suggest that our standard errors for the data have also been overestimated.

A way to evaluate the models without using the estimation of standard error of each datum is to apply the F-ratio test [Gordon et al., 1987]. The F-ratio test compares the variances but not the estimated standard errors of each datum. We follow the same formulation of the F-ratio test as Gordon et al. [1987] and obtain an F value of 1.33 when comparing the 3P model and the three BFP models. This value is less than the reference value of 2.76, corresponding to 3 versus 60 degrees of freedom at the 95% confidence interval. The number 3 equals the number of independent parameters of the 3 BFP models minus those of the 3P model (9 − 6), and 60 equals the number of degrees of freedom for the 3 BFP models. This means that the variances of the BFP and the 3P models are not significantly different, and thus, there is no statistical evidence for systematic error in the data or intraplate deformation.

C. Results

According to an instantaneous kinematic description of microplates, Schouten et al. [1989] predict that the two instantaneous Euler vectors for relative motions should lie at the propagating rift tips of the East Rift and West Rift. The locations of our Euler vectors for relative motion of EAS are closer to the rift tips than those of Engeln and Stein [1984] but are not at the rift tips (Fig. 11). This indicates that the motions of the Easter Microplate should be somewhat different than those predicted by the schematic model of Schouten et al. [1989]. This will be discussed in the sections on triple junction analysis and microplate rigidity.

The RM2 NAZ–PAC Euler vector, based on spreading rates from 0–3 Ma predicts faster rates than the 3P Euler vector (Fig. 7). This would suggest that spreading rates along the NAZ–PAC spreading axis have slowed down. However, a more recent global plate motion model NUVEL-1 [DeMets et al., 1989] is in closer agreement with our new

NAZ–PAC Euler vector. For the most part, the faster rates predicted by the RM2 appear to be related to the NAZ–PAC and PAC–Cocos rates used to derive RM2 [C. DeMets, pers. comm., 1988]. These rates are systematically faster than our rates and more recently estimated rates [DeMets et al., 1989; Gordon et al., 1988; Macdonald et al., 1988].

The NAZ–PAC differential vectors (3P NAZ–PAC vectors subtracted from NUVEL NAZ–PAC vectors) are of less magnitude and slightly different orientation than the PAC–HS differential vectors of Pollitz [1986]. This suggests that the recent ~3.2 Ma change in the Pacific plate absolute motion was not sudden [Pollitz, 1986] but instead was gradual and continued into the 0–3 Ma time period. Or else, the Nazca abso-

lute plate motion was not constant during that time period, although this possibility appears to be less likely because of the linear (hotspot) seamount chains on the Nazca plate (Fig. 12).

Absolute Motions

A. Methods

Angular vectors describing plate motions of the Nazca and Easter plates with respect to the Pacific hotspot frame of reference were obtained by adding the 3P Euler vectors for relative motion to the Pacific hotspot Euler vectors. Ideally, the 0–1 Ma Euler vector [Epp, 1984]

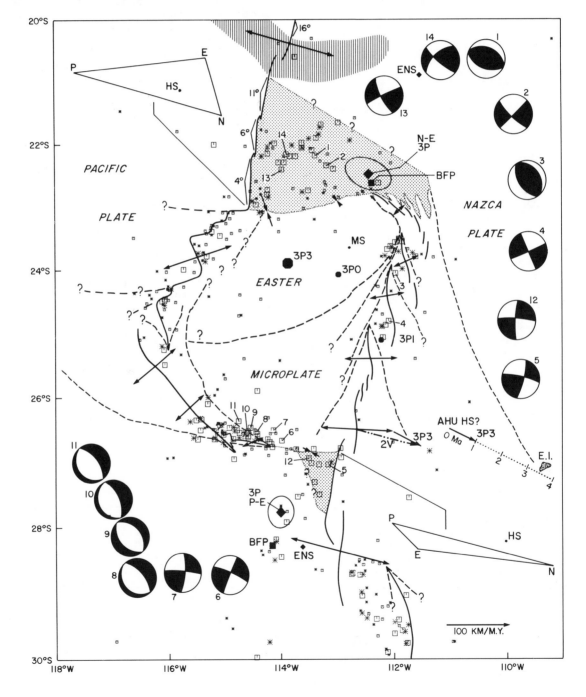

should be the best to use because it has about the time span as our relative Euler vector (0–0.73 Ma). But when it is added to our NAZ–PAC relative motion Euler vector, the NAZ absolute motion Euler vector does not predict absolute motion vectors that are parallel to hotspot traces on the Nazca plate. Thus, we also used two other starting Euler vectors for the absolute motion of the Pacific plate for the following two time periods: 0–3.2 Ma [Pollitz, 1986] and 0–10 Ma [Minster and Jordan, 1978]. We also compare the preferred Nazca absolute motion vector from Minster and Jordan [1978] (AM1-2; A12 in Fig. 10). We used the preferred Euler vector based solely on Pacific hotspot data [Pollitz, 1986] instead of the Euler vector resulting from a joint inversion of Pacific and North America hotspot data [Pollitz, 1988] because the former is for a younger time span (0–3.2 Ma) than the latter (0–9 Ma). Our choice is of minor consequence because the two Euler vectors are indistinguishable from each other at the 97.5% confidence level [Pollitz, 1988]. For the Galapagos area, we ignore the northwest-southeast trend of the 'second-type of hotspot islands' that have already been satisfactorily explained [Morgan, 1978].

The Easter absolute motion Euler vectors cannot be tested because there are no elongated seamount chains on the Easter microplate. However, two observations can be made. The 3P3 pole is the most centered with respect to the geometry of the microplate. The 3P0 pole is the closest to to the location of the Easter microplate's second-order moment pole (MS in Fig. 11) which is located along a line between the two relative motion poles of the microplate at a distance three and a half times closer to the NAZ-EAS pole than the PAC-EAS pole [Schouten et al., 1989]. The significance of these observations is unclear.

B. Assessment of the Absolute Motion Models

The AM1-2 NAZ absolute motion Euler vector is very near the 3P Euler vector for Nazca–Pacific motion (A12 and 3P in Fig. 10). This indicates that the orientation of near-axis seamount chains on the Nazca plate younger than 0.73 Ma that were previously attributed to NAZ absolute motions could actually be a result of recent NAZ–PAC relative motions.

Absolute motion vectors from the 3P1, 3P3, 3P0, and the A12 Euler vectors are shown with respect to seamounts and islands in areas of suspected hotspot traces (Fig. 12). The orientations of the 3P1 and 3P0 vectors do not match the trend of the Galapagos Islands (Fig. 12a), nor do they match the trend of the Juan Fernandez seamount chain (Fig. 12b). The A12 vector matches the Galapagos trend but not as well as the 3P3 vector does. The 3P3 and A12 vectors (Fig. 12b) match the trend of the Juan Fernandez seamount chain equally well but have different

predictions on the age progression of the seamounts which cannot be tested until more age information is obtained.

Comparing the four vectors along the Easter/Sala y Gomez seamount chain is much more tenuous. The origin of this chain has long been an enigma with respect to hotspot/plate tectonic theory [Wilson, 1963; Morgan, 1971; Duncan and Hargraves, 1984; Okal and Cazenave, 1985; Pilger and Handschumacher, 1981; Craig et al., 1984; Hey et al., 1985; Schilling et al., 1985; Hanan and Schilling, 1989; Baker et al., 1988; Hagen et al., 1989] and has had alternative origins proposed such as a 'hotline' [Bonatti et al., 1977] or an incipient rift [Mammerickx and Sandwell, 1986]. The difficulty in modeling this chain is that the distribution of seamounts and ridges is largely scattered, there may be more than one hotspot active, and/or there may be asthenospheric channeling to the East Rift from a hotspot somewhere to the east [Craig et al., 1984; Schilling et al., 1985] producing a 'second type' of hotspot trace [Morgan, 1978]. The orientation of this 'second type' of hotspot trace (2V in Fig. 11) represents the motion of the East Rift with respect to the hotspot frame.

Two proposed hotspot locations shown in Fig. 12c represent a recently active seamount ~190 km WNW of Easter Island at the western edge of the Ahu Volcanic Field [Baker et al., 1988; Hagen et al., 1989] and Sala y Gomez Island. The 3P1, 3P0, 3P3, and A12 absolute motion vectors are shown for each location. The entire Ahu Volcanic Field, which extends ~130 km towards Easter Island, appears to have been recently active based on SeaMARC II side-scan imagery [Hagen et al., 1989]. The general orientation of seamounts within this area is matched best by the 3P3 vector. Furthermore, the initial age of formation of Easter Island is predicted to be about 3.5 Ma by this vector [Fig. 11]. This age is in close agreement with the oldest subaerial rocks of Easter Island which have K-Ar ages ranging from about 2.5 to 3 Ma [Clark and Dymond, 1977; Baker et al., 1974].

The large scatter of seamount locations, the large data gaps east of Easter Island, and the uncertainty that Sala y Gomez Island represents the current location of an active hotspot, prevent detailed assessment of the vectors east of Sala y Gomez Island. However, based on the bathymetry available and the assumption that Sala y Gomez is the current location of an active hotspot, the 3P1 vector matches the main seamount trend, the 3P0 and A12 match another less significant trend, and the 3P3 vector does not match either trend, except within the immediate vicinity of Sala y Gomez Island (Fig. 12c).

C. Results

Of the four Euler vectors for Nazca absolute motion, the 3P3 Euler vector has the best fit to the orientation of seamounts near the Galapa-

Fig. 11. Tectonic boundaries, relative and absolute Euler poles, plate motion vectors, and two triple junction velocity triangles of the Easter microplate, with two types of earthquake epicenters and fault plane mechanisms [Engeln and Stein, 1984]. Large (Mb > 5) and small open squares represent NGDC earthquakes from 1963-1986. Large (Mb > 5) and small asterisks represent ISC earthquakes from 1972-1984. All solid lines between plates represent divergent boundaries except for transforms along the West Rift which are parallel to relative motion. Determination of the active West Rift boundaries was aided by GLORIA data [Rusby et al., 1988]. Short dashed lines represent pseudofaults. Long dashed lines represent fracture zones. The northern and southern most relative motion vectors are predicted from the 3P NAZ–PAC Euler vector (Fig 10.). The other vectors on the West Rift and East Rift are also predicted from the 3P PAC-EAS and NAZ-EAS Euler vectors (3P P-E and 3P N-E). Error ellipse about the 3P poles represents 95% confidence level (2-sigma). Our best fitting poles (BFP) for the respective plate pairs and the preferred Engeln and Stein [1984] Euler poles (ENS) are shown as solid squares and diamonds. Our preferred hotspot-EAS

the pole of second-order motion [Schouten et al., 1989] of the Easter microplate located 3.5 times closer to the NAZ–EAS than the PAC–EAS pole. Vertical line stipple near 21°S represents disturbed area of Rea [1978]. Large area of stipple represents deforming wedge of Nazca plate and the northern boundary of the Easter microplate. Its boundaries are based on earthquake epicentral pattern, bathymetry, magnetics, SeaMARC II side-scan, and orientation of the EPR (004°, 006°, Euler pole (3P3) is shown as a large solid octahedron. The other hotspot–EAS Euler poles are small solid octahedrons. MS represents 011°) in an area where it is predicted to be 016°. The southern stipple represent the uncertain location of the southern triple junction. HS represents the hotspot frame of reference for each velocity triangle. 2V vector marks the migration of the shallow East Rift summit with respect to the hotspot frame. The line extending from AHU HS predicts age of hotspot volcanism, which predicts that Easter Island was initially formed about 3.5 Ma. Other labels are same as those in Fig. 10. Lower-right vector represents magnitude scale.

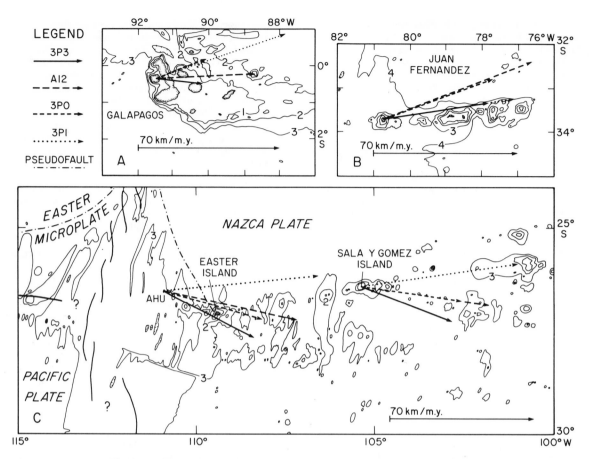

Fig. 12. Absolute motion vectors for three suspected hotspot seamount chains on Nazca plate and one-km isobaths [GEBCO, 1979]. For description of vectors see text and Fig. 11 caption. 3P3 vector (solid) is preferred in box A and B. 3P3 vector is preferred for Ahu Volcanic Field [Hagen et al., 1989] and Easter Island area in box C. No conclusion is drawn from the vectors at Sala y Gomez Island because of the sparse bathymetry data and the uncertainty that it is the current location of a hotspot. Active tectonic boundaries of Easter microplate shown in box C are same as in Fig. 1.

gos Islands, the Juan Fernandez Islands, and the Ahu Volcanic Field–Easter Island area. The A12 Euler vector has a lesser fit to these areas. The 3P0 Euler vector does not match the first two areas, and matches the third area to a lesser degree. The 3P1 vector does not match any of the trends (except, possibly, for the Sala y Gomez Island area), probably because of the uncertainty of absolute Pacific plate motion direction during its short 1 m.y. time span.

Triple Junction Analysis

The proposed northern triple junction near 23°03′S, 114°32′W analyzed in this section may not be the correct one or may not exist at all. For example, there may be another small microplate just north of the Easter microplate acting as a buffer (BUF) between the converging NAZ and EAS plates thus making it a PAC–EAS–BUF triple junction instead of a NAZ–PAC–EAS triple junction. Or the proposed triple junction may instead be a PAC–EAS ridge-transform intersection with an active transform to the west and a fracture zone to the east. Support for these two possibilities is that the earthquake epicenter pattern is diffuse and does not follow the northern boundary as one would ex-

pect if the northern boundary was indeed a NAZ–EAS or BUF–EAS transform fault. In addition, it must be kept in mind that assessing stability of this northern triple junction may be inappropriate because of the rapid plate boundary reorganization taking place because of the northward propagation of the East Rift. Unfortunately, due to the lack of a more promising location, it is analyzed for stability in the McKenzie and Morgan [1969] sense with the assumption that the relative motions predicted by the Euler vectors obtained in this study are correct.

The orientation of the predicted vector for NAZ–EAS transform motion is oblique to the orientation of the proposed NAZ–EAS transform fault (Fig. 11) which has been mapped as a shallow east-west trending ridge by Sea Beam [Hey et al., 1985; Francheteau et al., 1988]. Thus, the triple junction velocity triangle cannot be stable if this shallow ridge is considered a transform fault even though this structure is collinear with the PAC–EAS transform. It would be stable if the shallow ridge were a BUF–EAS transform fault with relative motion parallel to the ridge. But, as discussed later, it appears unlikely that a BUF plate exists. If the NAZ–EAS boundary is a convergent boundary and fixed to NAZ (i.e., all the deformation occurs on EAS) then the triple junction could not be stable and would require that the EPR north of the microplate propagate southward. This possibility can be ruled out because there is no evidence for southward propagation of the EPR in the Sea Beam bathymetry [Francheteau et al., 1988] nor in the Sea MARC II bathymetry and side-scan data [Naar et al., 1989].

However, the northern triple junction is stable if the NAZ–EAS boundary is a convergent boundary fixed to EAS instead of NAZ because it is collinear with the PAC–EAS transform fault (Fig. 11). The geomorphology of the northern structure [Naar et al., 1989] is consistent with convergent motion. The bathymetry is deeper to the south of the structure and there is a steeper slope on the south side than on the north side of the structure. There are no deep earthquake hypocenters in this region but there are shallow thrust-type fault plane mechanisms distributed mostly to the north of this boundary (within the large stippled area of Fig. 11). This suggests that all the convergent motion appears to be accommodated by crustal deformation to the north of this boundary and that this boundary is fixed to the microplate. The trend of the broad northern structure is collinear with the PAC–EAS transform near the triple junction [Francheteau et al., 1988; Naar et al., 1989]. Thus the conditions necessary for the stability of the northern triple junction are met if this structure is indeed a triple junction.

The exact location and geometry of the southern triple junction are still unclear, even after recently collected Sea Beam [Francheteau et al., 1987, 1988], unpublished MW8711 SeaMARC II data, and new GLORIA data [Rusby et al., 1988]. We analyze the proposed ridge-ridge-fault triple junction near 26°53′S 113°20′W [Francheteau et al., 1988] for stability. The predicted vector of PAC–EAS motion is oblique to the observed E–W structure proposed to be the PAC–EAS transform fault (Fig. 11). Thus the triple junction cannot be stable if the PAC–EAS boundary is considered a transform (Fig. 11). However, if the PAC–EAS boundary is considered a convergent boundary then the triple junction will be stable with one of the following: 7% asymmetry on the NAZ–PAC boundary (faster to the west), 8% asymmetry on the NAZ–EAS boundary (faster to the east), 15° counter-clockwise oblique spreading direction for NAZ–EAS, 18° clockwise oblique spreading direction for NAZ–PAC, or some combination of the above. Analysis of the magnetics, bathymetry, and SeaMARC II side-scan and GLORIA side-scan data [Rusby et al., 1988] indicate that plate boundaries in this region have been frequently reorganized by rift propagation. This could act as a mechanism to keep the triple junction stable. On the other hand, this also suggests that the concept of triple junction stability may also be inappropriate for this area.

Microplate Rigidity

Nearby Euler vectors predict the directions of relative plate motion to change rapidly along all microplate boundaries. This of course is not required if the plates are behaving non-rigidly, e.g., the pervasive simple shear model of Hey et al. [1986] and McKenzie [1986] for propagating rift systems, where spreading rates vary linearly from zero (at the tips of the propagating and failing rifts) to the full rate (at the point of no overlap). This model, by itself, cannot match all the recently collected structural data. Thus, we concur with previous work [Engeln and Stein, 1984; Hey et al., 1985; Engeln et al., 1988] that plate rotation is required.

Figure 11 shows that there may be oblique spreading on the proposed West Rift near 25.5°S and along the proposed East Rift near 23.5°S. Convergent motion is predicted along most of the proposed northern boundary and the part east of 114°W along the proposed southern boundary. The oblique compression at the southern boundary near the southern triple junction is 40 km/m.y. and the orthogonal convergence only amounts to 15 km/m.y. The maximum compression along the microplate boundaries is at the northern boundary just east of the proposed northern triple junction and has a total (orthogonal) convergence rate of about 50 km/m.y. Multiplying this rate by half of the ~200 km

distance from the triple junction to the NAZ–EAS Euler vector gives ~5000 square kilometers of lithosphere that should overlap every million years (assuming the NAZ–EAS Euler vector remains fixed).

How this convergent deformation is accommodated is another problem. Is it occurring at or near the boundaries? Is it being distributed over only the young part of the microplate or the young part of all three plates? Is it being distributed equally over the entire microplate? Is it being distributed as a function of plate age? Answers to these questions are being addressed [Naar et al., 1989] by analyzing the structural pattern of the microplate and its boundaries. Intraplate deformation has been proposed for the Gorda plate based on kinematic analyses [Stoddard, 1987; Wilson, 1986, 1988]. We suspect that some kind of intraplate deformation and rapidly migrating microplate boundaries are mostly responsible for the accommodation of this predicted fast convergence.

In terms of large-scale rift propagation the slow rate of profile 19 suggests the EPR to the north of the microplate is already starting to fail (slow down). This is supported by preliminary results [Sinton, pers. comm., 1988] that show the EPR between the 20.5°S offset and the microplate has a geochemical pattern similar to the Galapagos 95.5°W doomed-failing rift where basalts show a limited degree of fractionation (e.g., high Mg values) [Sinton et al. 1983; Yonover et al., 1986]. The orientations of the EPR segments between 20.5°S and 23°S are rotated counterclockwise with respect to segments farther north (Fig. 11). This pattern suggests the EPR between 20.5°S and 23°S is in a different stress regime (related to the northern propagating East Rift). Finally, the distribution of earthquakes north of the hypothesized northern boundary of the microplate, and the compressive and strike slip fault-plane mechanisms oblique to the observed structures (Fig. 11) also support the idea that the Nazca plate is deforming over a large area because of its interaction with the microplate.

It appears that the EPR just to the north of the microplate is slowing down in response to the northward propagating East Rift and thus has recently changed its spreading rate from NAZ–PAC rates to a slower rate. Thus, it appears that describing plate motions just to the north of the microplate with Euler vectors is inadequate, and that some kind of deformation must be occurring in response to the propagating East Rift. This implies that the northern triple junction is not located at a point and instead is a diffuse junction. The previous idea that there is a separate (buffer) plate in this northern area is difficult to substantiate because the earthquake epicenters cover most of the area and do not indicate a fourth aseismic plate. Furthermore, the abyssal hill fabric in this region (large stippled area in Fig. 11), imaged by SeaMARC II [Naar et al., 1989] and GLORIA [Rusby et al., 1988], is disrupted by cross-cutting faults and tilted blocks. In the eastern part of this region there is oblique abyssal hill fabric that cannot be attributed to rotation of the microplate but is clearly the result of some kind of broad transform shear.

Even though the majority of the microplate appears to be behaving mostly rigidly today, it must be kept in mind that the lithosphere transferred from the NAZ plate to the EAS microplate had to pass through this finite zone of transform deformation in some continuous fashion and had to have been deformed in the past. We suggest this transfer process is continuous because of the continuous oblique pattern of the eastern-most pseudofault (Fig. 11). Thus, we suspect the large-scale rift propagation in this area is characterized by a slow continuous process that has a broad deformation zone that migrates northward. It would shear and compress the plate fabric that is being transferred from the Nazca to the Easter plate because of the transform shear couple and the microplate rotation. To speculate, the East Rift probably starts to propagate northward once the convergent deformation of a certain area reaches a maximum amount.

Hey et al. [1985] proposed a ball-bearing analogy for the instantaneous motion of the microplate, with a rapid (15°/m.y.) rotation about a pole near the center of the microplate. Schouten et al. [1989] quantify this analogy with a second-order plate motion model. Their model does not attempt to describe the origin of the microplate, but predicts the present instantaneous plate motions. According to their model, pseudofaults on the interior of the microplate should form small circles about a pole (MS in Fig. 11) describing the microplate's second-order motion. The geometry of the two major (inner) pseudofaults on the microplate (Fig. 11) partially support this prediction. Their model assumes that rift propagation is only along preexisting transform boundaries and does not lengthen the East Rift (towards the north). The NNW trending eastern-most pseudofault clearly shows that the East Rift has been continually extending northward. The N–S rifted grabens in the Pito Deep area indicate that the East Rift is still attempting to propagate further to the north. Thus, some but not all of the predictions are satisfied by their model. This is not surprising, because the Euler vectors are not at the tips of the propagating rifts nor do the rifts only propagate to accommodate the rotation of the microplate as postulated in their model.

Conclusions

Instantaneous relative plate motions have been calculated for the Pacific, Easter and Nazca plates by inverting 69 data. The rates along the Nazca–Pacific and the Nazca–Easter spreading axes increase to the south. The rates along the Pacific–Easter spreading axis decrease to the south. The 0.49–0.50 Ma Emperor reversed event [Wilson and Hey, 1981] is evident in most of the profiles with fast rates. Nazca–Pacific transform faults do not currently exist along the superfast spreading portion of the EPR south of the Garrett transform fault. Instead, ridge offsets are accommodated by non-transforms (microplates, propagating rift systems, or overlapping spreading centers). The origin of the Easter microplate cannot be attributed solely to fast spreading rates along the preexisting Nazca–Pacific boundary because the fastest seafloor spreading exists to the south of the microplate. However, the fast rates combined with other factors such as the Easter hotspot or intensive mantle convection cannot be ruled out as being responsible for its origin.

The Nazca–Pacific Euler vector (0–0.73 Ma) from this study has a slower angular velocity and lies outside the confidence ellipse of the Minster and Jordan [1978] RM2 Euler vector (0–3.0 Ma). It also lies outside the confidence ellipse of the DeMets et al. [1989] NUVEL-1 Euler vector (0–3.0 Ma), but has approximately the same angular velocity. The Nazca–Easter Euler vector is just north of the northward propagating East Rift tip (Pito Deep). The Pacific–Easter Euler vector is south of the southeastward propagating West Rift. Our preferred Euler vector describing the absolute motion of the Easter microplate is the one derived for 0.0–3.2 Ma and is near the center of the microplate with an angular velocity of about 15°/m.y., making it a fast 'spinning' plate. The best fitting poles for all three plate pairs are similar to the poles derived by the closure condition which suggests the microplate is behaving mostly rigidly. Reduced chi-squared values and F-ratio tests support this finding. However, comparing the predicted motion vectors with the observed structures hypothesized to be microplate boundaries indicate that deformation must be occurring in some places. This could explain the compressive earthquake fault-plane solutions and the general epicentral pattern north of the northern boundary.

The proposed northern triple junction is unstable if it is treated as a ridge-fault-fault but can be stable as a ridge-fault-convergent triple junction if the convergent boundary is fixed to the Easter plate and remains collinear with the Pacific–Easter transform. This of course, assumes that the relative motion vectors predicted by our new Euler vectors at the triple junction are correct. However, several lines of evidence indicate that the proposed northern triple junction does not exist at just one point but instead is actually diffused over the area north of the microplate. The location and geometry of the southern triple junction is still unclear but has been tentatively proposed to be ridge-ridge-fault [Francheteau et al., 1988] which requires oblique convergence along the

Pacific–Easter transform and thus is unstable. However, if the southern triple junction is treated as a ridge-ridge-convergent triple junction it could be stable with a small amount of asymmetric or oblique spreading. It is not clear that a triple junction actually exists for any length of time in this area because the plate boundaries appear to be rapidly reorganizing. Discrepancies between observed plate boundaries and predicted motion vectors can be attributed to rapidly changing instantaneous Euler vectors during the Brunhes time interval, deformation near the boundaries, and/or intraplate deformation on one or more plates. Which of these processes predominates is uncertain but is suspected to be shear and compressive deformation localized in a band along the northern boundary of the microplate resulting from a combination of rift propagation and rapid microplate rotation.

Acknowledgments. Funding was provided by grants from the National Science Foundation (OCE82-15396, OCE84-10196 and OCE86-10412), the Office of Naval Research (to conduct wax modeling experiments), and the Whole Earth Society of Scripps Institution of Oceanography. We thank B. Minster and R. Parker for the use of their programs. We thank J. Francheteau, P. Patriat, S. Stein, R. Searle, J. Sinton, M. Kleinrock, D. Caress, J. Phipps Morgan, J. Sempere, N. Sleep, J. Engeln, C. DeMets, T. Reed, and M. Wilke for discussions. K.C. Macdonald provided SeaMARC II data from the Garrett and the Wilkes transform faults, prior to publication. J-G. Schilling did likewise with magnetic and 3.5 kHz data. For technical assistance at SIO we thank M. Keeler, A. Foster, and the Geologic Data Center. For technical assistance from HIG we thank D. Myhre, S. Stahl, S. McClintock, T. Duennebier, M. Simpson, B. Cunningham, J. Teo, J. Ahuna, C. Yoshinaga, S. Dang, B. Bays, and the SeaMARC II group. R. Hagen, A. Klaus, and P. Humphrey reviewed an earlier version of this manuscript. A special thanks goes to Wade Bartlett for his dedicated effort in mosaicking a tremendous amount of SeaMARC II data. We thank Captain Hayes, the crew of the R/V *Moana Wave,* and our watch standers. We are indebted to the inhabitants of Rapanui for their warm hospitality and thank the government of Chile for permission to work in the area. For very helpful reviews we thank Bernard Minster and Ken Macdonald. Hawaii Institute of Geophysics contribution no. 2179.

References

Abbott, D., A statistical correlation between ridge crest offsets and spreading rate, Geophys. Res. Lett., 13, 157–160, 1986.

Anderson, R.N., D.W. Forsyth, P. Molnar, and J. Mammerickx, Fault plane solutions of earthquakes on the Nazca plate boundaries and the Easter plate, Earth Planet. Sci. Lett., 24, 188–202, 1974.

Anderson, R.N., and J.G. Sclater, Topography and evolution of the East Pacific Rise between 5°S and 20°S, Earth Planet. Sci. Lett., 14, 433–441, 1972.

Baker, N.A., R.A. Hagen, D.F. Naar, and R.N. Hey, Bathymetry and submarine geology in the vicinity of Easter Island from SeaMARC II and Sea Beam data, EOS Trans. AGU, 69, 1429, 1988.

Baker, P.E., F. Buckley, and J.G. Holland, Petrology and geochemistry of Easter Island, Contr. Mineral. Petrol., 44, 85–100, 1974.

Bonatti, E., C.G.A. Harrison, D.E. Fisher, J. Honnorez, J-G. Schilling, J.J. Stipp and M. Sentilli, Easter volcanic chain (Southeast Pacific): A mantle hot line, J. Geophys. Res., 82, 2457–2578, 1977.

Champion, D.E., G.B. Dalrymple, and M.A. Kuntz, Radiometric and paleomagnetic evidence for the Emperor reversed polarity event at 0.46 ± 0.05 m.y. in basalt lava flows from the eastern Snake River Plain, Idaho, Geophys. Res. Lett., 8, 1055–1058, 1981.

Clark, J.G., and J. Dymond, Geochronology and petrochemistry of Easter and Sala y Gomez islands: Implications for the origin of the Sala y Gomez ridge, J. Volc. Geotherm. Res., 2, 29–48, 1977.

Craig, H., K.R. Kim, and W. Rison, Easter Island hotspot: I. Bathymetry, helium isotopes, and hydrothermal methane and helium, EOS Trans. AGU, 65, 1140, 1984.

DeMets, C., R.G. Gordon, D.F. Argus, and S. Stein, Current plate motions, submitted to the Geophys. Jour. of the Royal Astron. Soc., 1989.

Duncan, R.A., and R.B. Hargraves, Plate tectonic evolution of the Caribbean region in the mantle reference frame, In: The Caribbean–South American Plate Boundary and Regional Tectonics, W.E. Bonini, R.B. Hargraves and R. Shagam, eds., Geol. Soc. Am. Mem. 162, 81–93, 1984.

Dziewonski, A.M., T.-A. Chou, and J.H. Woodhouse, Determination of earthquake source parameters from waveform data for studies of global and regional seismicity, J. Geophys. Res., 86, 2825–2852, 1981.

Dziewonski, A.M., A. Fredman, D. Giardini, and J.H. Woodhouse, Global seismicity of 1982: centroid-moment tensor solutions for 308 earthquakes, Phys. Earth Planet. Int., 33, 76–90, 1983.

Dziewonski, A.M., J.E. Franzen, and J.H. Woodhouse, Centroid-moment tensor solutions for for July–September, 1983, Phys. Earth Planet. Int., 34, 1–8, 1984a.

Dziewonski, A.M., J.E. Franzen, and J.H. Woodhouse, Centroid-moment tensor solutions for for January–March, 1984, Phys. Earth Planet. Int., 34, 209–219, 1984b.

Dziewonski, A.M., J.E. Franzen, and J.H. Woodhouse, Centroid-moment tensor solutions for for October–December, 1984, Phys. Earth Planet. Int., 39, 147–156, 1985.

Dziewonski, A.M., G. Ekstrom, J.E. Franzen, and J.H. Woodhouse, Centroid-moment tensor solutions for for January–March, 1986, Phys. Earth Planet. Int., 45, 1–10, 1987a.

Dziewonski, A.M., G. Ekstrom, J.E. Franzen, and J.H. Woodhouse, Global seismicity of 1977: centroid-moment tensor solutions for 471 earthquakes, Phys. Earth Planet. Int., 45, 11–36, 1987b.

Dziewonski, A.M., G. Ekstrom, J.E. Franzen, and J.H. Woodhouse, Global seismicity of 1978: centroid-moment tensor solutions for 512 earthquakes, Phys. Earth Planet. Int., 46, 316–342, 1987c.

Dziewonski, A.M., G. Ekstrom, J.E. Franzen, and J.H. Woodhouse, Global seismicity of 1979: centroid-moment tensor solutions for 524 earthquakes, Phys. Earth Planet. Int., 48, 18–46, 1987d.

Engeln, J.F., and S. Stein, Tectonics of the Easter plate, Earth Planet. Sci. Lett., 68, 259–270, 1984.

Engeln, J.F., S. Stein, J. Werner, and R.G. Gordon, Microplate and shear zone models for oceanic spreading center reorganizations, J. Geophys. Res., 93, 2839–2856, 1988.

Epp, D., Possible Perturbations to hotspot traces and implications for the origin and structures of the Line Islands, J. Geophys. Res., 89, 11273–11286, 1984.

Forsyth, D.W., Mechanisms of earthquakes and plate motions in the East Pacific, Earth Planet. Sci. Lett., 17, 189–193, 1972.

Francheteau J., A. Yelles-Chaouche, and H. Craig, The Juan Fernandez microplate north of the Nazca-Pacific-Antarctic plate junction at 35°S, Earth Planet. Sci. Lett., 86, 253–268, 1987.

Francheteau J., P. Patriat, J. Segougin, R. Armijo, M. Doucoure, A. Yelles-Chaouche, J. Zukin, S. Calmant, D.F. Naar, and R.C. Searle, Pito and Orongo fracture zones: The northern and southern boundaries of the Easter microplate Southeast Pacific, Earth Planet. Sci. Lett., 89, 363–374, 1988.

General Bathymetric Chart of the Oceans, 5th ed., 1:10,000,000, Canadian Hydro. Serv., Ottawa, Sheet 5.11 SE Pacific Ocean, 1979.

Gordon, R.G., S. Stein, C. DeMets, and D.F. Argus, Statistical tests for closure of plate motion circuits, Geophys. Res. Lett., 14, 587–590, 1987.

Gordon, R.G., C. DeMets, D.F. Argus, and S. Stein, Current plate motions, EOS Trans. AGU, 69, 1416, 1988.

Hanan, B.B, and J-G. Schilling, Easter microplate evolution: Pb isotope evidence, J. Geophys. Res., 94, 7432–7448, 1989.

Handschumacher, D.W., R.H. Pilger, J.A. Foreman, and J.R. Campbell, Structure and evolution of the Easter plate, Geol. Soc. Am. Mem., 154, 63–76, 1981.

Hagen, R.A., N.A. Baker, D.F. Naar, and R.N. Hey, A SeaMARC II survey of recent submarine volcanism near Easter Island, submitted to Mar. Geophys. Res., 1989.

Herron, E.M., Two small crustal plates in the South Pacific near Easter Island, Nature Phys. Sci., 240, 35–37, 1972a.

Herron, E.M., Sea-floor spreading and the Cenozoic history of the East-central Pacific, Geol. Soc. Am. Bull., 83, 1671–1692, 1972b.

Hey, R.N., D.F. Naar, M.C. Kleinrock, W.J. Phipps Morgan, E. Morales, and J.-G. Schilling, Microplate tectonics along a superfast seafloor spreading system near Easter Island, Nature, 317, 320–325, 1985.

Hey, R.N., M.C. Kleinrock, S.P. Miller, T.M. Atwater, and R.C. Searle, Sea Beam/Deep-Tow investigation of an oceanic propagating rift system, J. Geophys. Res., 91, 3369–3393, 1986.

Hey, R.N., A. Klaus, W. Icay, and D.F. Naar, SeaMARC II survey of the propagating limb of a large non-transform offset along the fastest spreading EPR segment, EOS Trans. AGU, 69, 1429, 1988.

Kent, D.V., and F.M. Gradstein, A Jurassic to recent chronology; in Vogt, P.R., and B.E. Tucholke, eds., The Geology of North America, Volume M, The Western North Atlantic Region: Geological Society of America, chap. 3, 45–50, 1986.

Klaus, A., W. Icay, D.F. Naar, and R.N. Hey, SeaMARC II survey of a large non-transform offset along the fastest spreading East Pacific Rise segment, submitted to Mar. Geophys. Res., 1989.

Lonsdale, P.F., Overlapping rift zones at the 5.5 degree offset of the East Pacific Rise, J. Geophys. Res., 88, 9393–9406, 1983.

Lonsdale, P.F., Structural pattern of the Galapagos microplate and evolution of the Galapagos triple junctions, J. Geophys. Res., 93, 13551–13574, 1988.

Macdonald, K.C., and P.J. Fox, Overlapping spreading centers: A new kind of accretion geometry on the East Pacific Rise, Nature, 301, 55–58, 1983.

Macdonald, K.C., J.-C. Sempere, and P.J. Fox, Reply: The debate concerning overlapping spreading centers and mid-ocean ridge processes, J. Geophys. Res., 91, 10501–10511, 1986.

Macdonald, K.C., R.M. Haymon, S.P. Miller, J.-C. Sempere, and P.J. Fox, Deep-Tow and Sea Beam studies of dueling propagating ridges on the East Pacific Rise near 20°40′S, J. Geophys. Res., 93, 2875–2898, 1988.

Mammerickx, J., and D. Sandwell, Rifting of old oceanic lithosphere, J. Geophys. Res., 91, 1975–1988, 1986.

McKenzie, D.P., The geometry of propagating rifts, Earth Planet. Sci. Lett., 77, 176–186, 1986.

McKenzie, D.P., and W.J. Morgan, Evolution of triple junctions, Nature, 224, 125–133, 1969.

Minster, J.B., and T.H. Jordan, Present-day plate motions, J. Geophys. Res., 83, 5331–5354, 1978.

Minster, J.B., T.H. Jordan, P. Molnar, and E. Haines, Numerical modelling of instantaneous plate tectonics, Geophys. J. Roy. Astron. Soc., 36, 541–576, 1974.

Morgan, W.J., Convection plumes in the lower mantle, Nature, 230, 42–43, 1971.

Morgan, W.J., Rodriguez, Darwin, Amsterdam, .., a second type of hotspot island, J. Geophys. Res., 83, 5355–5360, 1978.

Naar, D.F., and R.N. Hey, Fast rift propagation along the East Pacific Rise near Easter Island, J. Geophys. Res., 91, 3425–3438, 1986.

Naar, D.F., and R.N. Hey, Speed limit for oceanic transform faults, Geology, 17, 420–422, 1989.

Naar, D.F., M.C. Kleinrock, R.N. Hey, D.W. Caress, K. Raeder, and D. Sandwell, Hot wax seafloor spreading experiments on video, EOS Trans. AGU, 67, 1228, 1986.

Naar, D.F., R.N. Hey, S. Stein, and T.B. Reed, Tectonic evolution of the

Easter Microplate: A SeaMARC II investigation, submitted to Mar. Geophys. Res., 1989.

Okal, E., and A. Cazenave, A model for the plate tectonic evolution of the east-central Pacific based on SEASAT investigations, Earth Planet. Sci. Lett., 72, 99–116, 1985.

Oldenburg, D.W., and J.N. Brune, Ridge transform fault spreading pattern in freezing wax, Science, 178, 301–304, 1972.

Oldenburg, D.W., and J.N. Brune, An explanation for the orthogonality of ocean ridges and transform faults, J. Geophys. Res., 80, 2575–2585, 1975.

Pilger, R.H., and D.W. Handschumacher, The fixed hotspot hypothesis and origin of the Easter-Sala y Gomez-Nazca trace, Geol. Soc. Am. Bull., 92, 437–446, 1981.

Pollitz, F.F., Pliocene change in Pacific-plate motion, Nature, 320, 738–741, 1986.

Pollitz, F.F., Episodic North America and Pacific plate motions, Tectonics, 7, 711–726, 1988.

Rea, D.K., Asymmetric sea-floor spreading and a nontransform axis offset: The East Pacific Rise 20°S survey area, Geol. Soc. Am. Bull., 89, 836–844, 1978.

Rea, D.K., and R.J. Blakely, Short-wavelength magnetic anomalies in a region of rapid seafloor spreading, Nature, 255, 126–128, 1975.

Rea, D.K., and K.F. Scheidegger, Eastern Pacific spreading rate fluctuation and its relation to Pacific area volcanic episodes, J. Volc. Geotherm. Res., 5, 135–148, 1979.

Rusby, R.H., R.C. Searle, J. Engeln, R.N. Hey, D. Naar, and J. Zukin, GLORIA and other surveys of the Easter and Juan Fernandez microplates, EOS Trans. Am. Geophys. Union, 69, 1428, 1988.

Ryan, W.B.F., Stratigraphy of late Quaternary sediments in the eastern Mediterranean, in D. J. Stanley, ed., The Mediterranean Sea: A natural sedimentation laboratory, Dowden, Hutchinson and Ross, Inc., Stroudsburg, 1972.

Sandwell, D.T., Thermal stress and the spacings of transform faults, J. Geophys. Res., 91, 6405–6417, 1986.

Schilling, J.-G., H. Sigurdsson, A.N. Davis, and R.N. Hey, Easter microplate evolution, Nature, 317, 325–331, 1985.

Schouten, H., K.D. Klitgord, and D.G. Gallo, Microplate kinematics of the second order, submitted to Earth Planet. Sci. Lett., 1989.

Searle, R.C., Multiple, closely spaced transform faults in fast-slipping fracture zones, Geology, 11, 607–610, 1983.

Sempere, J.-C., J. Gee, D.F. Naar, and R.N. Hey, Three-dimensional inversion of the magnetic field over the Nazca–Easter propagating rift near 25°S, 112°25′W, J. Geophys. Res., in press, 1989.

Sinton, J.M., D.S. Wilson, D.M. Christie, R.N. Hey, and J.R. Delaney, Petrological consequences of rift propagation on oceanic spreading ridges, Earth Planet. Sci. Lett., 62, 193–207, 1983.

Stoddard, P.R., A kinematic model for the evolution of the Gorda plate, J. Geophys. Res., 92, 11524–11532, 1987.

Talwani, M., C.C. Windisch, and M.G. Langseth, Reykjanes ridge crest: A detailed geophysical study, J. Geophys. Res., 76, 473–517, 1971.

Wilson, D.S., A kinematic model for the Gorda deformation zone as a diffuse southern boundary of the Juan de Fuca Plate, J. Geophys. Res., 91, 10529–10270, 1986.

Wilson, D.S., Tectonic history of the Juan de Fuca ridge over the last 40 million years, J. Geophys. Res., 93, 11863–11876, 1988.

Wilson, D.S., and R.N. Hey, The Galapagos axial magnetic anomaly: Evidence for the Emperor event within the Brunhes and for a two-layer magnetic source, Geophys. Res. Lett., 8, 1051–1054, 1981.

Wilson, J.T., A possible origin of the Hawaiian Islands, Can. J. Phys., 41, 863–870, 1963.

Woodhouse, J.H., and A.M. Dziewonski, Mapping the upper mantle: Three-dimensional modeling of Earth structure by inversion of seismic waveforms, J. Geophys. Res., 89, 5953–5986, 1984.

Yonover, R.N., J.M. Sinton, and D.M. Christie, ALVIN investigation of the petrological effects of rift failure, Galapagos spreading center near 95.5°W, EOS Trans. AGU, 67, 1185–1186, 1986.

KINEMATIC MODELS OF THE EVOLUTION OF THE GORDA RISE AND PRESIDENT JACKSON SEAMOUNT CHAIN

P. R. Stoddard

Department of Geological Sciences, Northwestern University, Evanston, Illinois

Abstract. The reorientation of the Juan de Fuca Plate at 5-10 Ma changed the geometry of the Gorda Rise and neighboring Blanco and Mendocino Transform Faults. The resulting ridge-transform-ridge geometry made transform-parallel, ridge-perpendicular spreading impossible, forcing segmentation and rotation of the Gorda Rise spreading center. This segmentation was accomplished by two propagators, and is reflected by both bathymetry and magnetic lineation pattern. A kinematic model is developed which accurately reproduces the magnetic lineations and is consistent with the bathymetric and dynamic constraints. In addition, two models are developed for the generation of the President Jackson Seamount chain. One calls for a hotspot, evidence of which may be found on both the Juan de Fuca and Pacific Plates; and the other, interaction of a weak hotspot, or "warm spot," with one of the propagator wakes resulting from the evolution of the Gorda Rise.

Introduction

The Gorda Rise is the southernmost segment of the Juan de Fuca ridge system, the locus of spreading between the Juan de Fuca and Pacific Plates (Figure 1). As this northern remnant of the Farallon Plate subducts, it has rotated, changing the geometry of the spreading regime of the Gorda Rise. This has resulted in a complex history of the rise from 5 Ma to present, as reflected in the magnetic lineation pattern it has produced.

Several methods by which spreading centers respond to changes in tectonic setting have been proposed and documented. Among these are: asymmetric spreading leading to rotation of ridge segments [Menard and Atwater, 1968; Stein et al., 1977], ridge jumps [Shih and Molnar, 1975; Winterer, 1976], and propagation of ridge segments [Hey, 1977; Hey et al., 1980; Hey and Wilson, 1982; Wilson et al., 1984; Johnson et al., 1983; Searle and Hey, 1983; Hey et al., 1986; Miller and Hey, 1986; Acton et al., 1987]. Features associated with such evolving ridges include overlapping spreading centers [Macdonald and Fox, 1983] small, non-overlapping offsets between spreading centers [Batiza and Margolis, 1986], and zero-offset transform faults [Schouten and White, 1980]. Evidence for ridge rotation [Menard and Atwater, 1968; Atwater and Menard, 1970] and propagation [Hey and Wilson, 1982] as well as overlapping of ridge segments of the Gorda Rise can be found in the magnetic and bathymetric data.

---

[1] Now at Department of Geology, Northern Illinois University, DeKalb, IL 60115

A kinematic model for the evolution of the Gorda Rise incorporating ridge rotation and propagation is developed. Additionally, two theories are proposed for the origin of a small chain of seamounts near the ridge; one is independent of the ridge evolution model, the other is a consequence of this model.

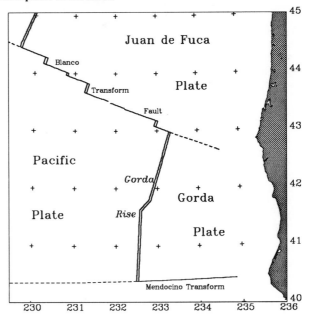

Fig. 1. Tectonic setting of the Gorda Rise.

Gorda Rise Evolution

Kinematic Constraints

Bathymetrically, the Gorda Rise can be seen as a broad high with a well-defined mid-axial depression, the Escanaba Trough [EEZ-SCAN, 1986; Figure 2]. For purposes of modeling, this trough is taken to represent the ridge position at present. The trough can be divided into at least three segments: north, middle, and south, with a small overlap of the northern and middle segments. The Blanco Transform is a region of highs and lows, small ridges, and pull-apart basins [Embley, 1987], while the Mendocino is marked by a narrow ridge. To the west of the Gorda Rise/Blanco Transform intersection lies an area of rugged topography, highlighted by the President Jackson Seamounts. The remainder of the region is relatively flat.

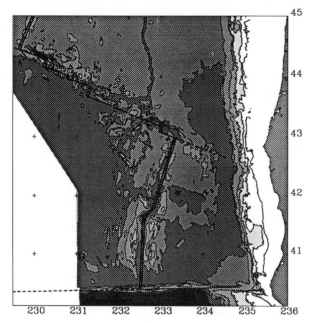

Fig. 2. Bathymetry of the Gorda Rise and environs. Depths range from >4000m (black) to <1000m (white). Contour interval is 500m.

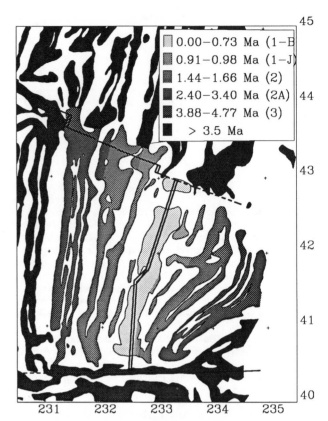

Fig. 3. Magnetic lineations produced at the Gorda Rise, as mapped by Raff and Mason (1961) and interpreted by Vine (1968). Anomaly ages are from Kent and Gradstein (1986). Anomaly numbers are in parentheses. 1-B is Brunhes normal. 1-J is Jaramillo normal.

Figure 3 shows the magnetic lineation pattern of the Gorda region as mapped by Raff and Mason [1961]. The interpretation is essentially that of Vine [1968]. The ages of anomalies range from 5.4 Ma for anomaly 3A to 0 Ma for anomaly 1 [Brunhes normal; Kent and Gradstein, 1986].

The most important features of this pattern, for the purposes of this study, are the clockwise rotation of the lineations and the disruptions in the magnetic signature on the Pacific Plate which may reflect propagator wakes (pseudo-faults) and/or hotspot activity. The trends of the middle portions of the lineations varies linearly from N5°E at anomaly 3A time (5.4 Ma) to N20°E at anomaly 1 (Jaramillo, 0.9 Ma), a change of about 3.5°/m.y. The northern portions of the lineations all trend approximately N18°E, parallel to the northern ridge segment. The placement of the northern propagator wake, which separates these two regions, is especially crucial in light of its proximity to the President Jackson Seamount chain. Points the wake should pass through include the offset in anomaly 3 (~43°N, 231.75°E), the offset in anomaly 2A (~42°N, 232°E), and the southern end of the northern ridge segment (~41.75°N, 232.75°E). Ideally, additional constraints on the wake location could be provided from the Gorda-side lineations; however, due to the subsequent deformation of these lineations, the only clear offset is that of anomaly 2A (Figure 3).

The evidence for a southern propagator is found by comparing bathymetric and magnetic data. The present trend of the southern ridge segment, as defined by bathymetry, is N6°E; while the trend of the corresponding portion of anomaly 1 (Jaramillo) is N20°E. Such a discrepancy in trends could be a result of either a dramatic ridge rotation in a counter-clockwise sense, which seems unlikely considering the overall clockwise rotation of anomalies, or by a propagating ridge.

Dynamic Constraints

The change of the Gorda Rise from a stable ridge segment to an actively evolving one was brought about by the change in the motion of the Juan de Fuca Plate at 5-10 Ma [Carlson, 1976; Nishimura et al.,

1984; Riddihough, 1984]. With this change in plate motion, the Blanco Transform Fault rotated about 25° clockwise, to a trend of ~N27°W while the trend of the Mendocino Transform, the southern boundary with the much older Pacific Plate, remained unchanged at N88°E. The resulting geometry, with the Blanco and Mendocino transforms no longer parallel to each other, make ridge-normal, transform-parallel spreading impossible along a linear Gorda Rise (Figure 4). The shading in Figure 4 shows relative age and, hence, relative strength of the lithosphere, from youngest and weakest (white regions) to oldest and strongest (dark regions). As the Gorda Rise is on some of the weakest lithosphere in this area, it is the part of the system that undergoes modification to allow transform-parallel spreading.

Ridge Evolution Model

Figure 5 shows, schematically, the proposed model of the Gorda Rise whereby the ridge is broken up into three segments. The northern and southern segments remain virtually perpendicular to their bounding transforms (the Blanco and Mendocino, respectively), thus allowing transform-parallel spreading. These two segments have zero initial length, but start propagating towards the center of the ridge, at the expense of the middle segment. Additionally, the new plate configuration creates a volume problem for the Gorda Plate, specifically, the line along which new material is being created (the Gorda Rise) is longer than the line along which old material is destroyed (the Cascadia subduction zone). As an apparent response to this problem, the northern segment is spreading faster than the southern, and therefore is

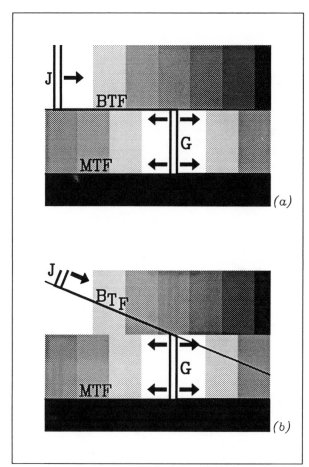

Fig. 4. Schematic diagram of the Juan de Fuca Ridge (J) - Blanco Transform (BTF) - Gorda Rise (G) - Mendocino Transform (MTF) system. Shading corresponds to age (and therefore strength) of lithosphere. Arrows indicate preferred direction of spreading. *(a)* Before the plate reconfiguration at 5 Ma. *(b)* After the plate reconfiguration. Note that ridge-perpendicular spreading is no longer consistent with transform-parallel spreading at the Gorda/Blanco intersection.

migrating eastwards with respect to the southern segment. The middle segment, whose southern end remains fixed to the southern segment, rotates clockwise, thus bridging the gap between the two propagators.

The following assumptions are made in this model. First, the segmentation does not begin until the rotation of the Juan de Fuca Plate is complete and the Blanco reaches its present trend, at an age of 5.3 Ma. Second, the Blanco Transform is treated as a simple transform fault. Recent work by Embley [1987] has shown that the Blanco actually consists of a series of strike-slip faults and pull-apart basins, but these should have no effect on the ridge evolution model.

Results

This simple model of the Gorda Rise fits the dynamic constraint of transform-parallel spreading, and the geometric constraints imposed by the observed magnetic lineation pattern and bathymetry. Figure 6 compares the lineation pattern predicted by the model with the observed pattern and shows the Gorda-side pattern that was produced by superimposing flexural-slip deformation [Stoddard, 1987] upon the Gorda lineations as produced by this model. As the figure shows, this model of

ridge evolution reproduces the Pacific pattern and, when incorporated with a suitable deformation model, can reproduce the Gorda lineations as well. The northern propagator wake breaks anomalies 3 and 2A (on both sides of the ridge) where there are breaks in the lineation pattern. The predicted final (present-day) position of the Gorda Rise coincides reasonably well with that of the bathymetric expression of the Escanaba Trough. The small offsets in the northern portion of the trough may be the result of small, recent breaks in the ridge. The spreading rates used are given in Table 1.

President Jackson Seamount Chain

Examination of the bathymetry in the region of the Gorda Rise (Figure 2) reveals a hummocky terrain immediately southwest of the Gorda Rise/Blanco Transform intersection, and a much smoother terrain further to the southwest. At or near the boundary between these two regions lies a series of at least four large, closely spaced volcanic seamounts on the Pacific Plate, called the President Jackson Seamounts (PJS), after the southeasternmost seamount in the chain. Two smaller Pacific seamounts, farther to the northwest but along the same trend, may also be part of the chain. Two possible origins of these seamounts are presented below.

Two-plate Hotspot Model

The trend of the PJS chain closely parallels the direction of absolute motion of the Pacific Plate, suggesting a hotspot, now located under President Jackson Seamount, as a possible origin of the seamounts. If such a hotspot exists, and if it has been long-lived enough, the chain should be evident on the Juan de Fuca Plate, north of the Blanco Transform Fault.

A model for such a two-plate hotspot track is diagrammed in Figure 7. The modeling was done using azimuthal equidistant projections centered about the point of interest at any given time. To find the Juan de Fuca portion of the hotspot track, the present hotspot location is first rotated back 2.5 Ma, until it intersects the transform fault (2.5p in Figure 7). This is accomplished using the Pacific/hotspot (pac/hs) pole of Engebretson et al. [1985], which is based on Minster and Jordan's [1978] pac/hs pole. Point M represents the rotated hotspot location using the Minster and Jordan [1978] pole. The error ellipse is the 67% confidence interval, assuming an arbitrary average error of 50 km in the six hotspot locations that went into the determination of this pole [Jurdy, 1988]. Next, using the Juan de Fuca/Pacific (jdf/pac) pole of Nishimura et al. [1984], the last 2.5 m.y. of right-lateral slip along the Blanco are accounted for, resulting in the hotspot position relative to the Juan de Fuca Plate at 2.5 Ma (2.5j in Figure 7). Addition of the pac/hs and jdf/pac vectors yields a Juan de Fuca/hotspot (jdf/hs) vector with a trend of ~N28°E and a rate of 29.4 mm/yr (vector S), which passes through two seamounts on the Juan de Fuca Plate.

The jdf/hs motion as determined here is compared with that predicted by previously published jdf/hs poles [Riddihough, 1984, vector R in Figure 7; Wilson et al., 1984, vector W]. The 67% confidence ellipse for vector S reflects only the errors associated with the jdf/pac rotation since the pac/hs motion is locally constrained by the model; however, those for vectors R and W reflect the combined pac/hs and jdf/pac rotations and are therefore much larger. The jdf/pac errors are estimated using the model of Stock and Molnar [1983] with a possible 10 km mislocation of the magnetic lineations used in the derivation of the rotation pole, and are then combined with the pac/hs errors using the error covariance matrices developed by Jurdy and Stefanick [1987]. Only the error ellipse of vector R [Riddihough, 1984] overlaps that of vector S; however, Wilson [1986] has adjusted his previous poles, so that they now more closely agree with Riddihough's [1984].

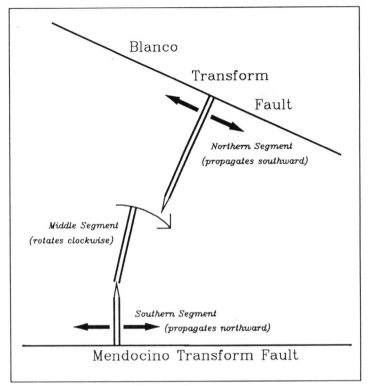

Fig. 5. Schematic diagram of evolution model of the Gorda Rise. Solid arrows indicate direction of spreading. Ridges pointed in direction of propagation.

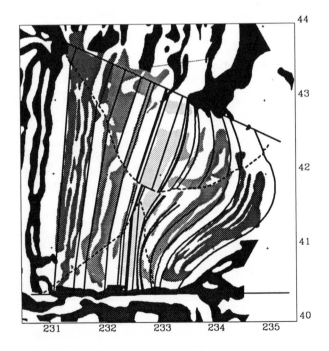

Fig. 6. Comparison of magnetic lineation pattern predicted by the model (heavy lines) with the observed pattern. Shading is the same as for Figure 4. Dashed lines indicate propagator wakes.

TABLE 1. Spreading Half-rates

Time, Ma	Northern rate, mm/yr	Southern rate, mm/yr
1.80-0.00	28.4	11.7
3.69-1.80	28.4	33.6
5.32-3.69	68.9	33.6

It must be stressed that the poles and points used in the above model were chosen to fit the two Juan de Fuca seamounts, with the additional constraint of the trend of the four Pacific seamounts. Obviously, the location of the hotspot at present and the choice of which segment of the Blanco Transform was active will greatly affect the results. The hotspot may actually lie farther southeast, while the hotspot/Blanco intersection may lie much farther northwest. An alternative model could be constructed that fits the volcanic outcrops which trend slightly more easterly than the vector R in Figure 7. These models are not intended to prove the existence of a long-lived hotspot, but to demonstrate that the volcanic trace on two plates is not inconsistent with a hotspot model.

"Warm Spot" Model

An alternative origin of the PJS chain is suggested by a comparison of the ridge evolution model with the absolute motion of the Pacific Plate. The northern Pacific-side propagator wake (dashed line), the absolute motion trend of the Pacific Plate (dotted line), and presumed present-day location of the PJS hotspot (star) are plotted against local bathymetry in Figure 8. It has been proposed that non-transform trending seamount chains may be related to ridge propagation [Lonsdale,

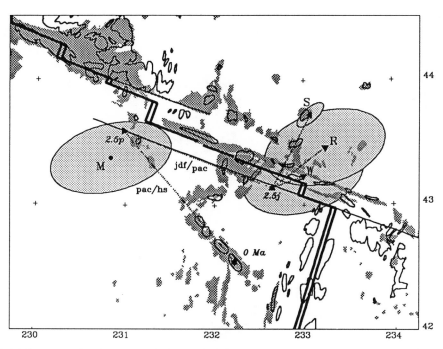

Fig. 7. Model for a two-plate hotspot track (dotted line) for a hotspot presently located under President Jackson Seamount (asterisk). *2.5p*: Position of hotspot relative to Pacific plate at 2.5 Ma (using pac/hs pole of Engebretson et al., 1985). M: Reconstructed position of asterisk using Minster and Jordan's (1978) pac/hs pole. *2.5j*: Position of hotspot relative to Juan de Fuca plate at 2.5 Ma. S: Juan de Fuca/hotspot motion (this study), from vector addition of pac/hs and jdf/pac vectors shown. R: JdF/hs motion from Riddihough (1984). W: JdF/hs motion from Wilson et al. (1984). M, S, R, and W shown with corresponding 67% confidence ellipses. Also shown are volcanic outcrops (dark shading) and 2500m contour for comparison.

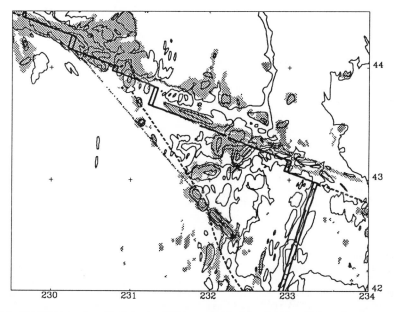

Fig. 8. Volcanism (shaded regions) and bathymetry (500m contours) of the President Jackson Seamount region. Asterisk shows assumed present-day position of the proposed hotspot. Dotted line shows motion of the hotspot with respect to the Pacific plate. Dashed line is the northern propagator wake. Note that these lines intersect near the first of the large seamounts in the President Jackson chain.

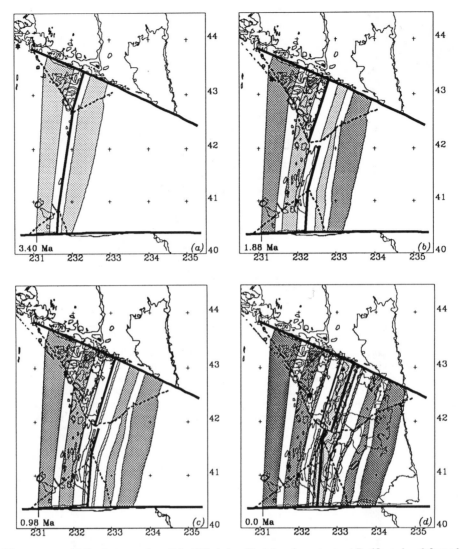

Fig. 9. Warm spot model for the generation of the PJS chain. Shaded regions represent Pacific and undeformed Gorda magnetic lineations, with the darker shades corresponding to older crust. Dashed lines are propagator wakes; dotted is the hotspot track. Also shown are the 3000m and 2500m bathymetric contours. *(a)* 3.40 Ma. *(b)* 1.88 Ma. *(c)* 0.98 Ma. Note that the hotspot track intersects the propagator wake at the northwestern end of the PJS chain. *(d)* Present.

1985; Schouten et al., 1987]; however, the PJS chain does not follow the propagator wake for lineations younger than anomaly 3 (3.88 Ma), so these models do not appear valid for the PJS chain. The intersection of the wake with the absolute motion line coincides with the northwestern end of the PJS chain, suggesting that the seamounts, even though they do not follow the propagator wake, are in some way influenced by it.

I therefore propose the following model for the generation of the President Jackson Seamount chain (Figure 9). A weak hotspot, or "warm spot," is present in the vicinity of the Blanco Transform (Figure 9a) at ~3.4 Ma. This warm spot may be the remnant of the passage of the Blanco Transform-Ridge system. At the time of Figure 9a (3.4 Ma) the warm spot is under ~3 m.y. old lithosphere, but is too weak a thermal anomaly to burn through the lithosphere. By 0.98 Ma (Figure 9c) the warm spot has reached the propagator wake, itself a possible zone of weakness, and begins to create the first of the large seamounts of the PJS chain. From 0.98 Ma to present, the warm spot has been under

increasingly younger (≤2.5 Ma) lithosphere, and has created the four large edifices of the chain.

This model is a variation of one developed by Davis and Karsten [1986]. Rather than a localized thermal anomaly, they propose that upwelling of chemical heterogeneities in the mantle may produce off-axis volcanism in anticipation of an overriding ridge. Again, passage of the Blanco Transform system over the heterogeneity may be the triggering mechanism.

Differentiation between these various alternatives will rely upon more detailed age data. The two-plate hotspot model predicts an age progression of seamounts along the proposed hotspot track on both the Pacific and Juan de Fuca Plates, while the warm spot theory makes no predictions about the Juan de Fuca seamounts. Unfortunately, dredge samples returned from the Pacific seamounts shows them to be very young MORBs in composition [D. Clague, personal communication], making accurate dating impossible at this time, and no samples have been returned from the two Juan de Fuca seamounts.

Summary

The segmentation of the Gorda Rise initiated with the change in motion of the Juan de Fuca Plate, and apparently has continued until the present. This segmentation appears to be a response to the new plate geometry, in an attempt to maintain transform-parallel spreading. The rise has reacted to the change in plate motion, rather than caused it.

Since the President Jackson Seamount chain does not follow the northern propagator wake, it does not appear to be related to the processes suggested by Lonsdale [1985] and Schouten et al. [1987]. The chain may be the result of a long-lived hotspot, as kinematic models can be devised consistent with seamount and volcanic trends on both the Pacific and Juan de Fuca Plates. Alternatively, volcanism from a weak hotspot, or "warm spot," may have been initiated by passage of the Blanco Transform-Ridge system over a chemical [Davis and Karsten, 1986] or thermal anomaly, resulting in the four Pacific Plate seamounts. Accurate dating of the Pacific and Juan de Fuca seamounts will be necessary to choose between these two models.

Acknowledgments. I thank Doug Wilson and Dick Hey for their reviews of this manuscript. I also thank Rodey Batiza and Jill Karsten for their helpful discussions, and Donna Jurdy for her suggestions on the manuscript. This work was supported by National Science Foundation grant EAR-8708558.

References

Acton, G. D., S. Stein, and J. F. Engeln, Formation of curved seafloor fabric by changes in rift propagation velocity and spreading rate: Application to the 95.5°W Galapagos propagator, *J. Geophys. Res.*, in press, 1987.

Atwater, T. M., and H. W. Menard, Magnetic lineations in the northeast Pacific, *Earth Planet. Sci. Lett., 7*, 445–450, 1970.

Batiza, R., and S. H. Margolis, Small non-overlapping offsets of the East Pacific Rise, *Nature, 320*, 439–441, 1986.

Carlson, R. L., Cenozoic plate convergence in the vicinity of the Pacific Northwest: A synthesis and assessment of plate tectonics in the northeastern Pacific, PhD thesis, University of Washington, Seattle, 130 pp., 1976.

Davis, E. E., and J. L. Karsten, On the cause of the asymmetric distribution of seamounts about the Juan de Fuca ridge: ridge-crest migration over a heterogeneous asthenosphere, *Earth Planet. Sci. Lett., 89*, 385–396, 1986.

EEZ-SCAN 84 Scientific staff, Atlas of the Exclusive Economic Zone, Western Conterminous United States, *U. S. Geol. Survey Misc. Invest.*, pp. 152, 1986.

Embley, R. W., Morphology, structure and resource potential of the Blanco Transform Fault Zone, *Geology and Resource Potential of the Continental Margin of Western North America and Adjacent Ocean Basins -- Beaufort Sea to Baja California: Circum-Pacifc Council for Energy and Mineral Resources Earth Science Series, VI*, Houston, TX, Circum-Pacific Council for Energy and Mineral Resources, 1987.

Engebretson, D. C., A. Cox, and R. G. Gordon, Relative motions between oceanic and continental plates in the Pacific basin, *Geol. Soc. Amer. Special Paper, 206*, 59 pp., 1985.

Hey, R., A new class of "pseudofaults" and their bearing on plate tectonics: A propagating rift model, *Earth Planet. Sci. Lett., 37*, 321–325, 1977.

Hey, R. N., and D. S. Wilson, Propagating rift explanation for the tectonic evolution of the Northeast Pacific -- the pseudomovie, *Earth Planet. Sci. Lett., 58*, 167–188, 1982.

Hey, R., F. K. Duennebier, and W. J. Morgan, Propagating rifts on mid-ocean ridges, *J. Geophys. Res., 85*, 3647–3658, 1980.

Hey, R. N., M. C. Kleinrock, S. P. Miller, T. M. Atwater, and R. C.

Searle, Sea Beam/Deep-Tow investigation of an active oceanic propagating rift system, Galapagos 95.5° W, *J. Geophys. Res., 91*, 3369–3393, 1986.

Johnson, H. P., J. L. Karsten, J. R. Delaney, E. E. Davis, R. G. Currie, and R. L. Chase, A detailed study of the Cobb offset of the Juan de Fuca Ridge: Evolution of a propagating rift, *J. Geophys. Res., 88*, 2297–2315, 1983.

Jurdy, D. M., Plate kinematic controls on accretion, *AGU Monograph Series*, in press, 1988.

Jurdy, D. M., and M. Stefanick, Errors in plate rotations as described by covariance matrices and their combination in reconstructions, *J. Geophys. Res., 92*, 6310–6318, 1987.

Kent, D. V., and F. M. Gradstein, A Jurassic to recent chronology, in *The Western North Atlantic Region, M*, edited by P. R. Vogt and B. E. Tucholke, pp. 45–50, The Geological Society of America, 1986.

Lonsdale, P., Nontransform offsets of the Pacific-Cocos plate boundary and their traces on the rise flank, *Geol. Soc. Am. Bull., 96*, 313–327, 1985.

Macdonald, K. C., and P. J. Fox, Overlapping spreading centres: New accretion geometry on the East Pacific Rise, *Nature, 302*, 55–58, 1983.

Menard, H. W., and T. Atwater, Changes in direction of sea-floor spreading, *Nature, 219*, 463–467, 1968.

Miller, S. P., and R. N. Hey, Three-dimensional magnetic modeling of a propagating rift, Galapagos 95°30'W, *J. Geophys. Res., 91*, 3395–3406, 1986.

Minster, J. B., and T. H. Jordan, Present-day plate motions, *J. Geophys. Res., 83*, 5331–5354, 1978.

Nishimura, C., D. S. Wilson, and R. N. Hey, Pole of rotation analysis of present-day Juan de Fuca plate motion, *J. Geophys. Res., 89*, 10,283–10,290, 1984.

Raff, A. D., and R. G. Mason, Magnetic survey off the west coast of North America, 40 N latitude to 50 N latitude, *Geol. Soc. Am. Bull., 72*, 1267–1270, 1961.

Riddihough, R., Recent movements of the Juan de Fuca plate system, *J. Geophys. Res., 89*, 6980–6994, 1984.

Schouten, H., and R. S. White, Zero-offset fracture zones, *Geology, 8*, 175–179, 1980.

Schouten, H., J. B. Dick, and K. D. Klitgord, Migration of mid-ocean ridge volcanic segments, *Nature, 326*, 835–839, 1987.

Searle, R. C., and R. N. Hey, Gloria observations of the propagating rift at 95.5°W on the Cocos-Nazca spreading center, *J. Geophys. Res., 88*, 6433–6448, 1983.

Shih, J., and P. Molnar, Analysis and implications of the sequence of ridge jumps that eliminated the Surveyor transform fault, *J. Geophys. Res., 80*, 4815–4822, 1975.

Stein, S., H. J. Melosh, and J. B. Minster, Ridge migration and asymmetric sea-floor spreading, *Earth Planet. Sci. Lett., 36*, 51–62, 1977.

Stock, J. M., and P. Molnar, Some geometrical aspects of uncertainties in combined plate reconstructions, *Geology, 11*, 697–701, 1983.

Stoddard, P. R., A kinematic model for the evolution of the Gorda plate, *J. Geophys. Res., 92*, 11,524–11,532, 1987.

Vine, F. J., Magnetic anomalies associated with mid-ocean ridges, in *The history of the earth's crust*, edited by R. A. Phinney, pp. 73–89, Princeton Univ. Press, Princeton, N. J., 1968.

Wilson, D. S., A kinematic model for the Gorda deformation zone as a diffuse southern boundary of the Juan de Fuca plate, *J. Geophys. Res., 91*, 10,259–10,269, 1986.

Wilson, D. S., R. N. Hey, and C. Nishimura, Propagation as a mechanism of reorientation of the Juan de Fuca Ridge, *J. Geophys. Res., 89*, 9215–9225, 1984.

Winterer, E. L., Anomalies in the tectonic evolution of the Pacific, in *The Geophysics of the Pacific Ocean basin and its Margin, 19*, edited by G. H. Sutton et al., pp. 269–280, AGU, Washington, D. C., 1976.

PHASE EQUILIBRIUM CONSTRAINTS ON THE EVOLUTION OF TRANSITIONAL AND MILDLY ALKALIC FE-TI BASALTS IN THE RIFT ZONES OF ICELAND

P. Thy

NASA, Johnson Space Center, SN2, Houston, Texas 77058

Abstract. Fe-Ti rich basalts from the Iceland and Galapagos propagating rifts show contrasting crystallization trends. Fe-Ti basalts occur in the western and south-eastern rift zones of Iceland and are plagioclase, olivine, and augite-phyric and span the range of compositions from quartz to mildly nepheline normative. Fe-Ti basalts from the Galapagos propagating rifts are quartz normative and are the product of advanced degrees of plagioclase, olivine, and augite cotectic crystallization relatively close to or at pigeonite saturation. Available melting experiments on tholeiitic to mildly alkalic basalts suggest that low-temperature liquids, coexisting with olivine, plagioclase, and augite, show a range in compositions which parallel the range observed in the natural Fe-Ti basalts. This range is mimicked by the augite compositions which become increasingly calcic going from tholeiitic, to transitional, and mildly alkalic basalts. The lack of correlation between normative composition and variables as $Mg/(Mg+Fe^{2+})$, TiO_2, and crystallization temperature suggests that the experimental data, as well as the natural basalts, cannot be related to a unifying liquid line of descent. Considering the good correspondence between low-temperature, experimental liquids and the natural Icelandic Fe-Ti basalts, the latter must represent high degrees of crystallization and can be explained by three-phase cotectic crystallization without reaching low-Ca pyroxene saturation. The Fe-Ti basalts of the Galapagos propagating rifts are quartz normative and crystallize pigeonite after about 40-50 % crystallization. These differences are consistent with phase equilibria of a model basalt phase diagram. Magmas parental to the Galapagos Fe-Ti basalts are quartz normative and plot to the oversaturated side while the parental magmas to the Icelandic Fe-Ti basalts are olivine normative and plot to the saturated side of the relevant plagioclase-augite-pigeonite join. The former may evolve past pseudo-peritectic relations towards quartz saturation while the latter will, dependent on the amount of crystal fractionation, tend to be consumed at pseudo-invariant, pigeonite saturated relations. Basaltic andesites and andesites are the products of Fe-Ti oxide crystallization. Because of the low degree of liquid left and the appearance of Fe-Ti oxides, the Icelandic basalts may never reach invariant relations. This is in contrast to the Galapagos propagating rift basalts where Fe-Ti oxides appear after pigeonite. These fundamental differences between the Iceland and the Galapagos propagating rifts reflect primary melt compositions, which can be related to deep-seated melting processes and the hotspot influence on normal ridge segments.

Introduction

Primitive oceanic glasses show a compositional range from quartz to mildly nepheline normative and span most of the range across the basalt tetrahedron of Yoder and Tilley [1962]. In addition to this first-order compositional variation, due to mantle processes, a second-order, low-pressure control by silicate phase equilibria and/or phenocryst redistribution parallels the primary variation and appears on many variation diagrams to be indistinguishable from the primary variation. Phenocryst populations suggest that olivine, plagioclase, and augite cotectics are the dominant control on low-pressure differentiation. Pigeonite is a rarely encountered phenocryst, but occurs in quartz normative Fe-Ti basalts of the Galapagos propagating rifts [Perfit and Fornari, 1983].

In terms of silicate equilibria, oceanic glasses, therefore, can analogously be described by the low-pressure, simplified anorthite (an)-diopside (di)-forsterite (fo)-quartz (q) tetrahedron of the $CaO-MgO-Al_2O_3-SiO_2$ system [Presnall et al., 1979]. The relevant cotectics in this quaternary are the fo-di-an univariant line and the enstatite saturated invariant point. A thermal divide between the tholeiitic and the alkalic systems is located relatively close to the fo-di-an join within the tholeiitic volume [Presnall et al., 1978]. Natural basaltic glasses often are displaced away from quaternary univariant relations [Presnall et al., 1979]. These displacements are caused by differences in the bulk systems and the non-quaternary nature of the natural systems. Part of this problem can be circumvented by determining cotectic relations in natural systems. Therefore, significant efforts have been put into the determination of phase boundaries in tholeiitic systems [Walker et al., 1979; Grove and Bryan, 1983; Tormey et al., 1987]. These experiments reveal an apparent, largely consistent position of the olivine (ol)-diopside (di)-plagioclase (pl) cotectic. Some experiments, furthermore, have pinpointed pigeonite saturated relations [Grove and Bryan, 1983]. The thermal divide between alkalic and transitional basalts are located slightly within the nepheline volume [Walker et al., 1979]. Because natural basalts are multicomponent systems, the ol-di-pl-q system is not quaternary and at the best a pseudo-system. For this reason, three-phase and four-phase saturated liquids cannot be expected to define lines and points, respectively. The subsequent text will refer to pseudo-univariant and pseudo-invariant relations or, when the meaning is clear from the context, simply univariant and invariant relations.

Studies of the natural phase relations in oceanic lavas often implicitly assume that mildly alkalic and transitional lavas can be understood in terms of tholeiitic phase equilibria. This appears to be supported by the work of Sack et al. [1987] which showed a continuous ol-pl-di cotectic bridging nepheline and low-Ca pyroxene saturated, pseudo-invariant relations of the alkalic and tholeiitic systems, respectively. Recent work by Thy [1988], nevertheless, suggests that the position of the liquid line of descent and the pseudo-univariant cotectic is a function of the bulk compositions. This means that, although the compositional variation in alkali-enriched tholeiites and mildly alkalic basalts may be controlled by olivine, augite, and plagioclase crystallization, distinct liquid lines of descent can be attributed to differences in normative composition of the parental magma. Because Fe-Ti enriched basalts represent high degrees of crystallization and may be relatively close to if not saturated in low-Ca pyroxene [Grove and Baker, 1984], such basalts potentially are useful for estimating the position of the pseudo-univariant cotectics and invariant relations as a function of bulk compositions.

This paper attempts to explore some of the silicate phase equilibrium constraints on the low-temperature and low-pressure evolution of mildly alkalic and transitional tholeiitic basalts from Iceland. Mildly alkalic and transitional Fe-Ti basalts of the recent rift zones of Iceland [Jakobsson, 1979; Thy, 1983a] are used as an example and are compared to tholeiitic basalts from the Galapagos propagating rifts [Fornari et al., 1983; Perfit and Fornari, 1983]. It is suggested that important phase equilibria constraints can be related to a range of primary melts.

Tectonic setting of the Fe-Ti basalts

Iceland straddles the mid-Atlantic ridge at approximately 65°N and consists of an extensive subaerial volcanic pile related to a ridge centered hotspot [Vogt, 1983]. Recent volcanic activity occurs along two main types of axial rift zones (Figure 1). The main tholeiitic rift zone is the landward extension of the mid-Atlantic ridge and is characterized by high heat flow, thin crust, and extensional tectonics [Jakobsson, 1972, 1980; Saemundsson, 1980; Oskarsson et al., 1982]. The western (Snaefellsnes) and the south-eastern (Sudurland-Vestmannaeyjar) flank zones (Figure 1), off the main rift zone, produce transitional and mildly alkalic olivine basalts [Jakobsson, 1979; Thy, 1983a; Meyer et al., 1985; Steinthorsson et al., 1985]. These flank zones are characterized by relatively low heat flow, a thicker crust, and poorly developed extensional tectonics. The volcanism is of Recent to Pliocene/Pleistocene ages and have, contrary to the main rift zone, no Tertiary equivalents [Saemundsson, 1980]. The flank zones, furthermore, are characterized by the development of topographically prominent volcanic edifices, in contrast to the graben dominated extension of the main rift zones.

Jakobsson [1979, 1980] identified spatial groupings of eruption sites active within a limited time interval and with comparable tectonic, petrographic, and chemical characteristics and referred to these as volcanic systems. The south-eastern flank zone (Sudurland-Vestmannaeyjar; Figure 2)

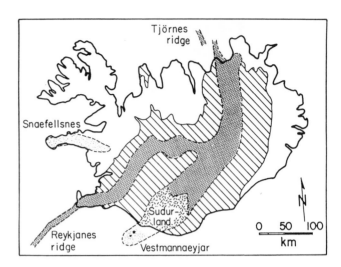

Fig. 1. Regional geology and petrology of Iceland after Jakobsson [1972] and Saemundsson [1980]. The recent rift zones (<0.7 m.y.; fine shaded areas) are flanked first by Pliocene-Pleistocene lavas (0.7-3.1 m.y.; ruled area), and then by the Tertiary series (>3.1 m.y.; white area). The Snaefellsnes rift zone produces mainly mildly alkalic lavas. The main rift zone, extending from the Tjörnes ridge in the north to the Reykjanes ridge towards the south, produces tholeiitic lavas. The south-eastern rift zone show a transition from the tholeiitic rift zone, to transitional alkalic lavas in Sudurland, and mildly alkalic lavas in the Vestmannaeyjar.

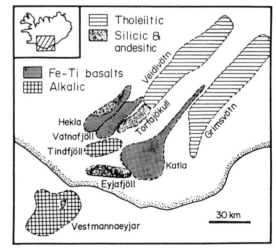

Fig. 2. Summary of the volcanic systems and petrographic characteristics of the south-eastern flank zone of Iceland. See Jakobsson [1979] for details and discussions.

contains nine such volcanic systems each showing distinct chemical features which can be related to their distance from the main rift zone [Jakobsson, 1979]. Going south, the Grimsvötn and Vedivötn systems are tholeiitic, the Hekla, Vatnafjöll, and Katla systems are characterized by Fe-Ti basalts, the Tindfjöll system produces transitional alkali basalts, and the southern-most center, the Vestmannaeyjar, produces mildly alkalic basalts. Intermediate and silicic lavas dominate the Torfajökull and Eyjafjöll systems and the recent eruption products from the Hekla volcano [Baldridge et al., 1973; Jakobsson, 1979]. The transition from tholeiitic, to Fe-Ti basalts, and alkalic lavas is considered to be the geochemical manifestation of a propagating rift system [Christie and Sinton, 1981]. For this reason, the south-eastern flank zone has been suggested to be a propagating rift [Meyer et al., 1985] or a transgressive rift zone [Oskarsson et al., 1982]. The western flank zone (Snaefellsnes, Figure 3) is oblique to the main rift zone and may represent a zone of volcanic systems which was separated and isolated during an eastward shift of the spreading axis [Johannesson, 1980; Oskarsson et al., 1982]. The Snaefellsnes zone is characterized by WNW-ESE fissures which produce mildly alkalic basalts to Fe-Ti basalts. The Snaefellsnes central volcano is composed of relatively silicic lavas [Steinthorsson et al., 1985]. Although the Snaefellsnes zone possesses some of the chemical features of a propagating rift [Steinthorsson et al., 1985], the available tectonic and chemical information are not conclusive.

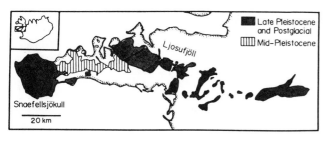

Fig. 3. Geology of the Snaefellsnes flank zone of western Iceland [after Saemundsson, 1980].

The Galapagos spreading center of the eastern Pacific Ocean separates the Cocos plate to the north and the Nazca plate to the south. The ridge is towards the west terminated by the Pacific plate at a triple junction and towards the east by the Panama fracture zone. The spreading center has been interpreted as formed by two propagating rifts moving, respectively, to the east (85°W) and to the west (95°W). Part of the spreading system has tectonically and chemically been influenced by the Galapagos Islands hotspot located slightly to the south [Schilling et al., 1982a; Fisk et al., 1982]. Tholeiitic Fe-Ti basalts and associated silicic rocks are produced relatively close to the tips of the propagating rifts [Christie and Sinton, 1981, 1986]. Sinton et al. [1983] and Sinton and Christie [1981] suggested that these basalts formed by extensive fractional crystallization in large isolated magma chambers reflecting perturbed supply and thermal regimes along the normally steady-state rift.

Chemistry and petrology of the Fe-Ti basalts

Major element analyses of Fe-Ti basalts from the flank zones of Iceland were obtained from Jakobsson [1979] and Thy [unpublished]. These include most of the major occurrences of recent Fe-Ti basalts: The volcanic centers Hekla, Vatnafjöll, and Katla of the south-eastern rift zone and areas around the Snaefellsnes central volcano of the Snaefellsnes flank zone. Other occurrences are the Skagi (2-0.5 m.y. old)

flank zone [Sigurdsson et al., 1978], a single dredged sample from the submarine Reykjanes ridge [Melson and O'Hearn, 1986], and segregation veins on the islands of Heimaey and Surtsey of the Vestmannaeyjar volcanic center [Thy and Jakobsson, unpublished]. Except for a few of the Snaefellsnes analyses, all are bulk rock XRF-analyses. For comparative purposes are included a series of microprobe glass analyses of tholeiitic basalts, Fe-Ti basalts, andesites, and dacites from the Galapagos spreading center of the eastern Pacific Ocean [Fornari et al., 1983; Melson et al., 1977]. The Icelandic andesites are, except for five basaltic andesites from Jakobsson [1979], microprobe glass analyses. Most of these analyses, shown as shaded fields on the figures, are on tephra from the Hekla volcano and are taken from unpublished sources. Discussions of the general petrology and geochemistry of the Icelandic flank zones are outside the scope of this paper. Detailed discussions can be found in Jakobsson [1979], Thy [1983a], Imsland [1983], Steinthorsson et al. [1985], Meyer et al. [1985], and many others.

The chemical analyses are shown on Figures 4 and 5 and representative analyses are given in Table 1. The four Icelandic groups of Fe-Ti basalts appear to be largely similar in terms of FeO-MgO-TiO$_2$ concentrations (Figure 4, Table 1). However, distinct differences between the volcanic systems can be observed when, in particular, K$_2$O and Na$_2$O are considered (Figure 5). The Vatnafjöll and Hekla Fe-Ti basalts have the lowest sodium and potassium contents, the Katla Fe-Ti basalts occupy an intermediate position, and the Snaefellsnes Fe-Ti basalts have the highest values. In addition, the same sequence shows an increase in K$_2$O/Na$_2$O ratios (Figure 5). This potassic nature of the Snaefellsnes Fe-Ti basalts reflects a well documented regional difference between the south-eastern and the western flank zones

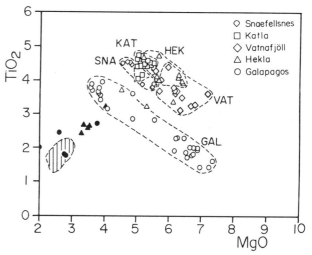

Fig. 4. TiO$_2$ versus MgO (wt.%) for Fe-Ti basalts from the Iceland and Galapagos spreading centers. KAT Katla; HEK Hekla; VAT Vatnafjöll; SNA Snaefellsnes; GAL Galapagos. The Galapagos data range from basalts to Fe-Ti basalts and are microprobe analyses of glasses. Filled symbols are basaltic andesites, andesites, and dacites. Shaded area represent andesitic glasses from the Hekla volcano. All analyses are calculated anhydrous with all iron as FeO. See text for sources to data.

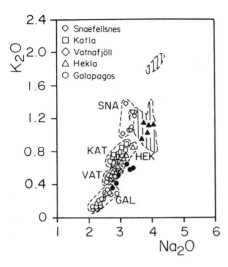

Fig. 5. K_2O versus Na_2O (wt.%) for Fe-Ti basalts from the Iceland and Galapagos spreading centers. Same data as for Figure 4.

[Jakobsson, 1972, 1980; Oskarsson et al., 1982; Thy, 1983a; Imsland, 1983; Steinthorsson et al., 1985]. The Galapagos Fe-Ti basalts show compositions clearly distinct from the Icelandic basalts. They have the lowest Na_2O, K_2O, and TiO_2 values of the lavas considered here and reach much higher FeO and lower MgO contents (Figures 4 and 5, Table 1). The andesitic lavas from the Galapagos and Iceland centers can clearly be separated from each other and both groups appear to be compositionally related to the associated Fe-Ti basalts.

The dominant phenocrysts of the Icelandic Fe-Ti basalts are plagioclase, olivine, and augite [Jakobsson, 1979; Thy, 1983a; Meyer et al., 1985; Steinthorsson et al., 1985]. In addi-

TABLE 1. Fe-Ti basalts from the Icelandic and
Galapagos spreading centers

	HEK	VAT	KAT	SNA	GAL
SiO_2	47.29	46.32	47.22	45.62	51.21
TiO_2	3.98	3.69	4.47	3.81	3.76
Al_2O_3	13.66	14.09	12.91	15.08	10.45
FeO	14.77	15.20	14.98	13.63	18.28
MnO	0.24	0.23	0.21	0.23	0.29
MgO	5.67	6.08	5.10	5.35	3.74
CaO	9.59	9.81	9.92	10.19	8.70
Na_2O	2.95	2.81	2.98	3.25	2.61
K_2O	0.62	0.53	0.76	1.05	0.30
P_2O_5	0.56	0.46	0.53	0.74	0.44
Total	99.30	99.22	99.08	98.95	99.78
$Mg/(Mg+Fe^{2+})$	0.44	0.45	0.41	0.45	0.30
$q/(q+pl+ol)$	0.21	0.14	0.20	-0.03	0.54

Average Fe-Ti basalts from the Hekla (HEK), Vatnafjöll (VAT), and Katla (KAT) volcanic centers from Jakobsson [1979, table 17]. The Fe-Ti basalt from the Snaefellsnes flank zone (SNA) is an average of analyses on scoria collected north-west and south of the Snaefellsnes central volcano. The Galapagos average Fe-Ti basalt (GAL) is calculated from the analyses given by Fornari et al. [1983].
$Mg/(Mg+Fe^{2+})$ is calculated with $Fe^{2+}/(Fe^{2+}+Fe^{3+})=0.86$.
$q/(q+pl+ol)$ is a normative ratio, with nepheline as negative q, and based on a molecular CIPW with $Fe^{2+}/(Fe^{2+}+Fe^{3+})=0.86$.

tion, subordinate amounts of titanomagnetite appear in lavas with FeO above 14 wt.% and TiO_2 above 4 wt.% [Thy, 1983b]. Jakobsson [1979] suggested that olivine, plagioclase, and augite crystallized largely simultaneously. The basaltic andesites and andesites of the Hekla volcano contain plagioclase, olivine, augite, and titanomagnetite [Sigvaldason, 1974; Baldridge et al., 1973]. Low-Ca pyroxenes have not been found in any of the Icelandic Fe-Ti basalts or associated andesitic lavas. The Fe-Ti basalts of the Galapagos propagating rifts contain plagioclase, augite, olivine, pigeonite, and minor amounts of titanomagnetite phenocrysts with pigeonite replacing olivine in the most evolved lavas [Perfit and Fornari, 1983]. The Galapagos andesites contain plagioclase, augite, pigeonite, and titanomagnetite.

Liquid lines of descent

The normative mineralogy of the basalts was calculated using a CIPW molecular norm with total iron distributed between Fe^{2+} and Fe^{3+} such that $Fe^{2+}/(Fe^{2+}+Fe^{3+})=0.86$ [Presnall et al., 1979]. The normative components were, following Presnall et al. [1979], reduced to the pseudo-quaternary ol-pl-di-q, recalculating hypersthene to equal amounts of olivine and quartz and representing nepheline (ne) by negative quartz. This pseudo-quaternary is shown in Figure 6 as projections on the triangular diagrams ol-di-q and ol-pl-q. The Vatnafjöll Fe-Ti basalts show the lowest normative diopside and plagioclase, followed by the Hekla lavas with intermediate values, and subsequently the Katla and Galapagos lavas with the highest values. The Snaefellsnes Fe-Ti basalts show a bimodality in normative diopside. The samples plotting with low diopside (and plagioclase) are whole rock analyses and those with high diopside are the matrix glass in the same samples.

Several interesting features appear on the projections (Figure 6). First, the Fe-Ti basalts occupy a broad band across the projections from quartz to nepheline normative compositions. Second, the individual volcanic centers are clearly discriminated from each other particular in terms of normative diopside and to a lesser degree also plagioclase. Third, the transition from Fe-Ti basalts to basaltic andesites for the Hekla and Galapagos series occurs at different levels of silica saturation.

As most of the Fe-Ti basalts contain plagioclase, augite, and olivine phenocrysts, it can be assumed that the dominant control on their evolution was pseudo-univariant, cotectic crystallization. Some of the Galapagos Fe-Ti basalts contain, in addition, a low-Ca pyroxene (pigeonite) and mark the pseudo-invariant cotectic where pigeonite replaces olivine. The strong linear trends, therefore, may be controlled by three-phase cotectic crystallization, which will project as lines in the pseudo-quaternary projections. However, two features remain to be explained: 1) The subparallel displacements in normative diopside and plagioclase and 2) the fact that individual centers apparently evolve towards distinct Fe-Ti basalt and andesite compositions. The first of these can be explained as an effect of pressure on the cotectics, which is known to expand the augite field and to a lesser degree the plagioclase field. It is interesting, if this interpretation is correct, that the Snaefellsnes whole-rock basalts may reflect higher pressure crystallization relative to the presumably low pressure matrix glass in the same samples (Figure 6). The second feature is unexpected and is inconsistent with the simple view that oceanic basalts evolve towards identical low-pressure,

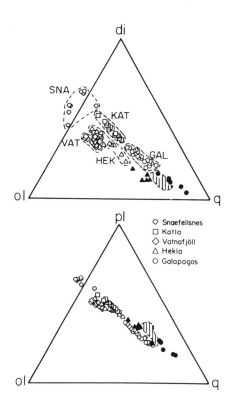

di

SNA
KAT
VAT
HEK
GAL

ol   q

pl

○ Snæfellsnes
□ Katla
◇ Vatnafjöll
△ Hekla
○ Galapagos

ol   q

Fig. 6. Normative projections of Fe-Ti basalts onto the triangular olivine (ol)-diopside (di)-quartz (q) and ol-plagioclase (pl)-q diagrams of the natural basaltic di-pl-ol-q pseudo-quaternary. Normative projections and data reductions are after Presnall et al. [1979]. Same data as shown on Figure 4.

pseudo-invariant relations and that andesites are related to augite, low-Ca pyroxene, and plagioclase cotectics.

It is generally accepted that Fe-Ti basalts form by extensive crystal fractionation in high level magma chambers. This has been demonstrated for the Galapagos spreading center by Perfit and Fornari [1983] and Clague et al. [1981] suggesting 40-70 per cent crystallization of plagioclase, augite, and olivine from a primitive mid-ocean ridge basalt parent in order to produce the most Fe and Ti enriched lavas observed. Crystallization temperatures for the pigeonite saturated glasses can be estimated to 1180-1100°C [Perfit and Fornari, 1983] which is consistent with a minimum of 1170°C estimated by Fisk et al. [1982] for pigeonite-free glasses. Using Rayleigh fractionation of P_2O_5, which has a bulk distribution coefficient approaching zero, most of the Icelandic Fe-Ti basalts can be estimated to represent 55-65 per cent crystallization. These estimates are highly uncertain because, although the most evolved Fe-Ti basalts generally contain 0.46-0.76 wt.% P_2O_5, the parent magma compositions are unknown and can only be inferred from nearby volcanic centers. Furthermore, the estimates probably represent minimum values as Thy [1988] experimentally demonstrated that the most primitive magma erupted from the mildly alkalic Surtsey volcano, at the southernmost tip of the Vestmannaeyjar, itself may reflects some unknown degree of crystallization. Despite these uncertainties, the Iceland and the Galapagos Fe-Ti basalts

appear to represent approximately equal amounts of crystallization. The minimum crystallization temperatures for the Iceland Fe-Ti basalts have been estimated to ~1130°C [Jakobsson, 1979; Thy, 1983a]. The significant lower crystallization temperature for the Icelandic Fe-Ti basalts and the comparable degrees of crystallization suggest that these can be expected to evolve towards pseudo-invariant relations at much lower temperatures than observed for the Galapagos Fe-Ti basalts. Potentially this, together with the highly diverse normative compositions, imply that the position of invariant relations may be shifting.

The Hekla Fe-Ti basalts contain the same silicate phenocryst assemblages as the associated basaltic andesites and andesites (olivine, augite, and plagioclase). This contrasts to the Galapagos lavas where the basalts contain plagioclase, augite, pigeonite, and olivine while the associated andesitic lavas contain plagioclase, augite, and pigeonite. In addition, magnetite is present in the most evolved Fe-Ti basalts as well as in the basaltic andesites of both the Galapagos and Iceland series. The compositional effect of magnetite crystallization is decreasing TiO_2 and FeO and strongly increasing normative quartz (or SiO_2) (Figures 4 and 6). It is evident from the variation diagrams that magnetite is a major factor in determining the formation of the basaltic andesites and andesites of the Hekla volcano and the Galapagos spreading rifts [cf., Clague et al., 1981]. In addition, it is also clear from the phenocrysts assemblages that the Hekla andesites evolve along a different cotectic than the Galapagos andesites.

Experimental constraints

In order to evaluate the evolution of the natural Fe-Ti basalts, one atmosphere, experimental pseudo-univariant liquids coexisting with olivine, augite, and plagioclase and pseudo-invariant liquids coexisting with pigeonite, olivine, augite, and plagioclase have been compiled from the literature. These form two groups of experimental starting compositions: tholeiitic and mildly alkalic. The tholeiitic group of experiments include Oceanographer fracture zone basalts [OFZ; Walker et al., 1979], Kane fracture zone basalts [KFZ; Tormey et al., 1987], a basalt from 43°N of the mid-Atlantic ridge [Grove and Bryan, 1983], a basalt from 23°N of the mid-Atlantic ridge [Grove and Bryan, 1983], and Mt. Pluto, FAMOUS basalts [Grove and Bryan, 1983]. The mildly alkalic group includes basalts from Pantelleria [Mahood and Baker, 1986] and the Surtsey volcano of the south-eastern volcanic zone of Iceland [Thy, 1988]. In addition was included liquids produced by heating stage crystallization of glass inclusions in plagioclase phenocrysts of a tholeiitic basalts from the Asal volcanic chain, Djibouti Republic [Clocchiatti and Massare, 1985].

When plotted on the normative projections (Figure 7) most of the experimental, pseudo-univariant tholeiitic liquids fall on a well defined line. This line, in general, can be represented by the liquid line of descent for the OFZ experiments as determined by Walker et al. [1979]. The pigeonite saturated liquids for the FAMOUS and 23°N experiments show a consistent position and terminate the univariant line at temperatures of about 1170°C. The OFZ liquids appear to extend the univariant line to relatively higher normative quartz and lower temperatures of 1080°C. This, however, may be an effect of Fe-Ti oxide crystallization in the low-Ca pyroxene saturated experiment reported by Walker et al. [1979]. The KFZ univariant glasses did not reach pigeonite

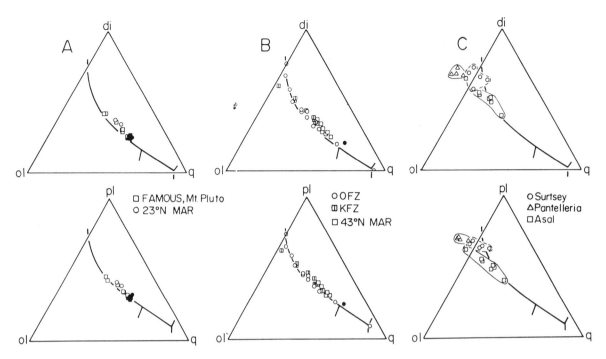

Fig. 7. One atmosphere, experimental pseudo-univariant liquids coexisting with olivine, plagioclase, and augite and pseudo-invariant liquids coexisting with pigeonite, olivine, augite, and plagioclase projected onto the triangular diagrams ol-di-q and ol-pl-q. Norm calculation and projections are after Presnall et al. [1979]. Heavy drawn curve is Walkers et al.'s [1979] liquid line of descent (see *B*). (*A*) FAMOUS and 23°N MAR depleted tholeiites. FAMOUS, Mt. Pluto, is after Grove and Bryan [1983]. 23°N, mid-Atlantic ridge, is after Grove and Bryan [1983]. Filled symbols mark pseudo-invariant liquids with pigeonite, olivine, augite, and plagioclase. (*B*) Oceanographer fracture zone (OFZ), Kane fracture zone (KFZ), and mid-Atlantic ridge at 43°N enriched tholeiites. OFZ is from Walker et al. [1979]. The filled symbol marks a glass which, in addition to plagioclase, olivine, and augite, coexist with an unspecified low-Ca pyroxene and Fe-Ti oxides. KFZ is based on Tormey et al. [1987]. 43°N MAR is from Grove and Bryan [1983]. (*C*) Surtsey and Pantelleria mildly alkalic basalts and Asal tholeiites. Surtsey volcano, Vestmannaeyjar, is based on Thy [1988]. Pantelleria, Strait of Sicily, is from Mahood and Baker [1986]. Asal is from Clocchiatti and Massare [1985] and is based on heating stage crystallization of glass inclusions in plagioclase phenocrysts in lavas from the Asal volcanic chain, Djibouti Republic.

crystallization and also plot with relatively lower normative quartz despite the fact that they reach approximately the same degree of crystallization (~50 %) and slightly lower temperature (~1150°C) than the FAMOUS experiments. Surprisingly, the 43°N experiments did not reach low-Ca pyroxene saturation although they reach approximately similar compositions and lower temperatures (1140°C) compared to the FAMOUS and 23°N liquids (Figures 7-9). The Asal liquids show compositional similarities to the KFZ liquids and appear, together with these, to straighting-out the high temperature segments defined by the OFZ experiments. The pseudo-univariant, mildly alkalic Surtsey liquids plot, as expected, at lower normative quartz (or higher nepheline) without reaching invariant relations and represent relatively high degrees of crystallization (~70 %) and low temperatures of ~1110°C. The Pantelleria pseudo-univariant (Fe-Ti oxide-free) liquids represent relatively low degrees (~40 %) of crystallization, but may be located at the undersaturated side of the thermal divide, while the Surtsey liquids may be on the saturated side. This observation, which is supported by phase equilibria considerations, as discussed below, places the thermal divide slightly to the undersaturated side of the ol-pl-di plane.

Viewing the experimental data as a function of TiO_2 and $Mg/(Mg+Fe^{2+})$ versus normative q/(ol+pl+q), three groups of pseudo-univariant liquids can be identified (Figures 8 and 9), and related to a spectrum of alkali enrichment/depletion or normative quartz (Figure 10). These groupings are substantiated by the pyroxene-liquid relations as shown on the normative ol-di-q projection (Figure 11).

First, the mildly alkalic liquids plot with relatively high TiO_2 and low $Mg/(Mg+Fe^{2+})$ and may reach slightly quartz normative compositions (positive q/(ol+pl+q)). The Surtsey pseudo-univariant liquids represent 60-70 per cent crystallization and straddle the ol-di-pl plane. Their coexisting augites are slightly nepheline normative and highly calcic and plot relatively close to normative diopside (Figure 11F). The Pantelleria liquids (25-40 % crystallized) coexist with augites plotting in the same general field as the Surtsey augites. Because the tie-lines for the Pantelleria liquids are rotated away from the quartz side of the di-ol-pl join, the liquids must be located to the undersaturated side of the thermal

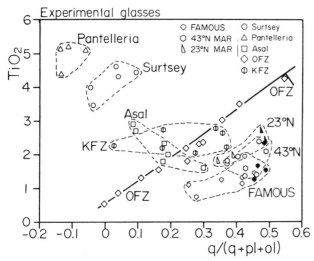

Fig. 8. TiO$_2$ versus normative q/(q+pl+ol) of one atmosphere, experimental liquids coexisting with either a pseudo-univariant olivine, augite, and plagioclase assemblage (open symbols) or a pseudo-invariant pigeonite, olivine, augite, and plagioclase assemblage (filled symbols). Norm calculation and data as for Figure 7. The norm calculation is a molecular CIPW norm with iron distributed according to Fe^{2+}/(Fe^{2+}+Fe^{3+})=0.86 and recasting hypersthene into quartz and olivine. Normative nepheline is shown as negative quartz.

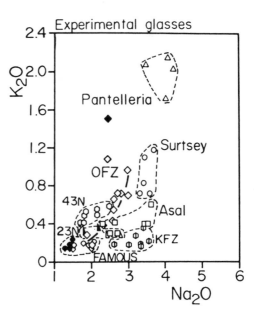

Fig. 10. K$_2$O versus Na$_2$O of one atmosphere, experimental liquids coexisting with either a pseudo-univariant olivine, augite, and plagioclase assemblage (open symbols) or a pseudo-invariant pigeonite, olivine, augite, and plagioclase assemblage (filled symbols). Data as for Figures 7 and 8.

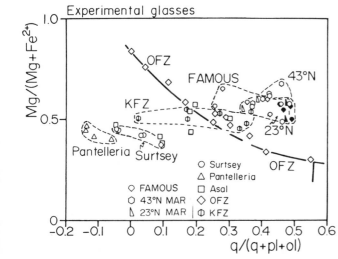

Fig. 9. Mg/(Mg+Fe^{2+}) versus normative q/(q+pl+ol) of one atmosphere, experimental liquids coexisting with either a pseudo-univariant olivine, augite, and plagioclase assemblage (open symbols) or a pseudo-invariant pigeonite, olivine, augite, and plagioclase assemblage (filled symbols). Norm calculation, data, and data reduction as for Figures 7 and 8.

divide between the tholeiitic and the alkalic volumes. On the other hand, the Surtsey liquids appear to be located to the saturated side of the thermal divide.

Second, the depleted tholeiitic liquids from FAMOUS, 23°N, and 43°N reach a much higher Mg/(Mg+Fe^{2+}), lower

TiO$_2$, and higher q/(ol+pl+q) and, except for the 43°N liquids, appear to reach pigeonite saturated relations. The augites coexisting with the depleted tholeiite liquids are relatively subcalcic (Figure 11). The pigeonite saturated experiments from FAMOUS and 23°N show nearly identical phase compositions on the normative projections (Figure 11A,B) but represent relatively wide ranges of crystallization: 50-65 % in the FAMOUS experiments and 20-55 % in the 23°N experiments. Therefore, the liquid compositions show wide ranges in FeO, MgO, and TiO$_2$ [Grove and Bryan, 1983]. These large crystallization intervals can be due to pseudo-invariant crystallization and the non-quaternary nature of the natural basalt system. It is clear that the FAMOUS and 23°N experiments reflect pigeonite saturated relations at relative low degrees of crystallization. The range in the degree of crystallization (20-50 %) when pigeonite first appear can be related to small differences in the experimental starting compositions.

Third, the KFZ (and Asal) enriched tholeiites occupy an intermediate position. The KFZ pseudo-univariant liquids reflect 45-55 per cent crystallization without reaching pigeonite saturation (Figure 11D). The main characteristics of this group appear to be the slightly less quartz normative compositions and the crystallization of augites compositional intermediate between those coexisting with the alkalic and the depleted tholeiite liquids. An effect of these displacements is a steepening of the tie-lines between liquids and augites (Figure 11D).

The OFZ pseudo-univariant liquids deserve special attention as these have been widely used as a model for the evolution of the mid-ocean ridge type of basalts. When the coexisting OFZ liquids and augites are plotted (Figure 11E), it is clearly revealed that the OFZ "liquid line of descent" is a composite of experiments on two different compositions

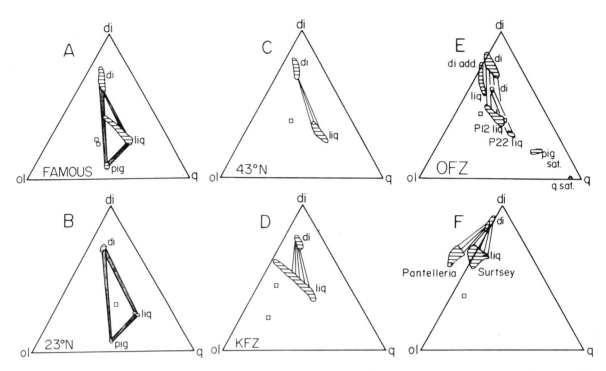

Fig. 11. Liquid-pyroxene relations for one atmosphere, experimental liquids projected on the normative ol-di-q triangular diagram. Norm calculations and data reduction as for Figure 7. Super-liquidus liquids are shown as squares. (A) FAMOUS, Mt. Pluto [Grove and Bryan, 1983]. Pseudo-univariant and invariant liquids and coexisting augites and pigeonites. Shaded area delineate tie-lines between coexisting augite, pigeonite, and liquid. (B) 23°N, mid-Atlantic ridge [Grove and Bryan, 1983]. Pseudo-invariant liquids and coexisting augites and pigeonites. Tie-lines are represented by the shaded areas. (C) 43°N, mid-Atlantic ridge [Grove and Bryan, 1983]. Pseudo-univariant liquids and coexisting augites which show crossing tie-lines. Sample P22 shows on Figure 11E a FAMOUS and 23°N (depleted tholeiite) type of descent while sample P12 reveals a KFZ (or enriched tholeiite) type of descent. Only the P22 sample can be expected to crystallize low-Ca pyroxene after relatively low degrees of crystallization while the P12 sample even after fairly high degrees of crystallization (68 % in the experiments) would not have reached invariant relations. Unfortunately, the low-Ca pyroxene crystallizing from the P22 sample were not fully characterized by Walker et al. [1979]. The high temperature OFZ liquids were produced by adding (together with plagioclase and olivine) diopside to the experimental charges. Although the composition of this diopside is not given, it is clear that tie-lines between this and the coexisting liquids would be crossing both the P22 and the P12 experimental tie-lines. For these reasons, the OFZ "liquid line of descent" is composed of three independent segments and should not be used as a unifying liquid line of descent for mid-ocean ridge basalts. The composite nature of the OFZ liquid line of descent may, in part, also explain why the OFZ experiments show well defined trends on most variation diagrams, while with tie-lines. (D) Kane fracture zone (KFZ) [Tormey et al., 1987]. Pseudo-univariant liquids and coexisting augites with representative tie-lines. (E) Oceanographer fracture zone (OFZ) [Walker et al., 1979]. Pseudo-univariant liquids and coexisting augites with tie-lines for samples P22 and P12. The most quartz rich P22 experiments are pseudo-invariant saturated containing unanalysed low-Ca and high-Ca pyroxenes (pig sat) and quartz (q sat). The location of additional experiments are indicated (di add, q sat). (F) Pantelleria and Surtsey [Mahood and Baker, 1986; Thy, 1988]. Pseudo-univariant liquids and coexisting augites are shown.

other experiments on mid-ocean ridge type of compositions (e.g., KFZ and FAMOUS) show much less correlation between important variables such as $Mg/(Mg+Fe^{2+})$ and $q/(ol+pl+q)$ (Figure 9). If the OFZ experiments are "broken down" into their individual components these can, in part, be understood in terms of the results obtained on other basalt types.

The fundamental phase relations of basaltic systems can qualitatively be described by the low-pressure simplified $CaO-MgO-Al_2O_3-SiO_2$ phase diagram (Figure 12). Typically primary mid-ocean ridge basalts, as well as transitional and mildly alkalic basalts, plotting within the olivine volume to the undersaturated side of the an-di-en join (A in Figure 12) can, by cooling-controlled equilibrium crystallization, be expected to be fully crystallized at the peritectic (p) point in equilibrium with forsterite, diopside, low-Ca pyroxene, and plagioclase. Only by a process of crystal fractionation can the liquid cross the peritectic point and evolve towards quartz saturation at the eutectic (e) point. An alternative evolution path is shown by quartz normative tholeiites (liquid B) which, plotting to the oversaturated side of the join an-di-en, would

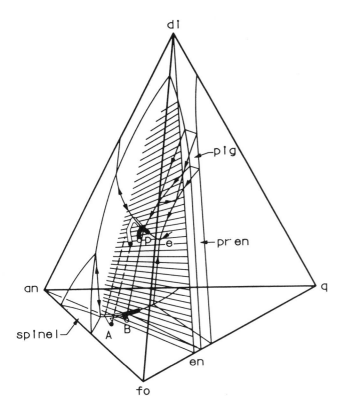

Fig. 12. The CaO-MgO-Al$_2$O$_3$-SiO$_2$ simplified basalt system after Presnall et al. [1979]. The crystallization of the liquid A plotting to the saturated side and the liquid B plotting to the oversaturated side of the join an-di-en are discussed in the text.

be able to cross, or circumvent, the peritectic point and evolve towards the eutectic (e) point. These basic constraints can be applied to natural systems.

The liquid fraction remaining at low-Ca pyroxene, pseudo-invariant relations will increase as a function of the composition of the parental magma going from mildly alkalic, to transitional, and depleted tholeiites. However, this increase will to a certain degree be compensated by the shifting augite compositions. The transitional groups and some mildly alkalic basalts will reach invariant relations at high degrees of crystallization and will there be fully consumed. Because of the very low fraction of liquid remaining for these basalts, invariant relations can experimentally and analytically not directly be determined. The depleted tholeiites will reach invariant relations with a relatively high fraction of liquid remaining and there, dependent on the starting composition, either be fully consumed or be able to cross the invariant and evolve towards "andesitic" compositions. This will explain the relatively high degrees of crystallization for the mildly alkalic and transitional basalts without reaching invariant relations in contrast to the lower degrees reached by the tholeiite group of experiments.

The starting compositions (or super-liquidus glasses) for the discussed experiments are shown on Figure 11. It is evident from these plots that only the OFZ P22 and 23°N experiments would be able to pass pseudo-invariant relations

evolving towards andesitic compositions (~80 % liquid remains when the 23°N experiments reach pigeonite saturation). The FAMOUS starting compositions plot close to the augite-pigeonite-plagioclase plane and may, if uncertainties in the treatment of Fe^{3+} and Fe^{2+} of the liquid and the augites are taken into consideration, evolve past pseudo-peritectic relations (~50 % liquid remains at the peritectic). The starting compositions for the 43°N, OFZ P12, KFZ, and Surtsey experiments plot, in the given sequence, increasingly away from the augite-pigeonite-plagioclase plane, and, therefore, can be expected to reach pseudo-invariant relations with decreasing liquid remaining.

The experimental data, thus, imply that the liquid line of descent is dependent on the silica saturation of the parental compositions. The general movement of equilibrium liquids away from quartz is paralleled by the clinopyroxene compositions which move towards normative diopside and slightly nepheline normative compositions. Shown on pseudo-quaternary projections most univariant liquids plot along a narrow band ranging from mildly nepheline to quartz normative compositions. However, each compositional group will evolve along distinct paths, in particular, for non-quaternary elements and may reach invariant relations at highly diverse crystallization temperatures. The depleted tholeiites reach pseudo-invariant relations at a temperature of 1170°C. In contrast, the enriched tholeiite and the mildly alkalic groups still at temperatures of 1150-1140°C and 1110°C, respectively, will not have reached pseudo-invariant relations which presumably must be located at much lower temperatures.

Origin of the Fe-Ti basalts

The phase relations and chemical variation of the natural Fe-Ti basalts can be understood in the light of the experimental constraints. The Icelandic Fe-Ti basalts evolve towards nearly identical TiO$_2$ levels and Mg/(Mg+Fe^{2+}) ratios, but show wide ranges in normative quartz (Figures 13 and 14)

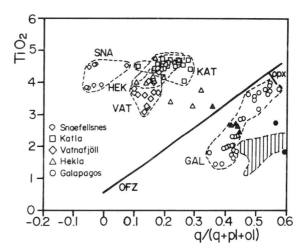

Fig. 13. TiO$_2$ versus normative q/(q+pl+ol) of Icelandic and Galapagos spreading center Fe-Ti basalts. Same data as for Figure 4 and calculations as for Figure 8. The Oceanographer fracture zone liquid line of descent (OFZ) is the same as shown on Figure 8. "opx" indicate the appearance of low-Ca pyroxene.

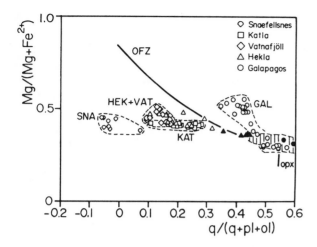

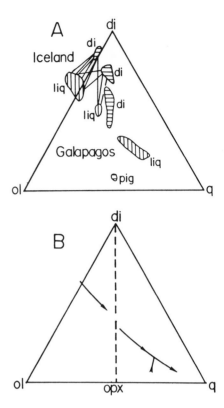

Fig. 14. $Mg/(Mg+Fe^{2+})$ versus normative $q/(q+pl+ol)$ of Icelandic and Galapagos spreading center Fe-Ti basalts. Norm calculation and data reduction as for Figure 9. Data shown are the same as for Figure 4. The Oceanographer fracture zone (OFZ) liquid line of descent is the same as shown on Figure 9. "opx" indicate the income of low-Ca pyroxene.

Fig. 15. (*A*) Pyroxene-liquid relations in natural basaltic glasses from the Galapagos and the Icelandic spreading centers. The Iceland data are transitional and mildly alkalic glasses taken from Thy [1983a]. The Galapagos data are from Fornari et al. [1983] and Perfit and Fornari [1983]. (*B*) Interpretation of the liquid lines of descent for mildly alkalic basalts and depleted tholeiites. The location and nature of mildly alkalic pseudo-invariant relations it not known.

and appear to reflect a continuum of liquid lines of descent bracket by those determined on enriched tholeiites and mildly alkalic olivine basalts, respectively. The available data on natural coexisting augites and liquids (Figure 15) show a general rotation of tie-lines, relatively to tholeiitic augites, away from normative quartz similar to those experimentally determined. Transitional basalts and mildly alkalic basalts, located to the saturated side of the thermal divide, will evolve towards increasing normative quartz (Katla, Hekla, and Vatnafjöll). These types of basalts crystallize augite chemically distinct from those found in the alkalic lavas and will therefore move along a different path. Some Icelandic mildly alkalic basalts [Thy, 1983a] are straddling the thermal divide with a similar normative composition as their evolved differentiation products (e.g., Snaefellsnes). These compositions may be "trapped" by pseudo-ternary invariant relations on the thermal divide.

The Galapagos spreading center basalts evolve towards quartz normative Fe-Ti basalts with significant lower $Mg/(Mg+Fe^{2+})$ ratios (Figure 14) and lower TiO_2 (Figure 13) than the Iceland basalts. The Galapagos liquid line of descent can be compared to the experimentally determined liquid line of descent for OFZ sample P22, but is clearly different from the depleted FAMOUS tholeiites (Figure 11). Although the compositions of coexisting liquid and pyroxenes for the Galapagos glasses are not known, the available data [Perfit and Fornari, 1983; Fornari et al., 1983] stress the similarities with the OFZ P22 experiments (Figures 11 and 15). The augites crystallizing from the Galapagos liquids are relatively subcalcic and clearly differ from the mildly alkalic and transitional augites (Figure 15).

The highly diverse compositional variation of the Fe-Ti basalts points to distinct cotectics or liquid lines of descent. Considering their relatively high degree of crystallization, the Fe-Ti basalts must be close to pseudo-invariant relations. This is clearly shown by the Galapagos glasses which, in addition to the three-phase cotectic assemblage, occasionally

contain a fourth phase (pigeonite) reflecting invariant relations. Pseudo-invariant relations for the Icelandic basalts have not been experimentally reproduced and, therefore, can not be located. However, it is unlikely that its location should coincide with that determined for the depleted tholeiites.

Some mildly alkalic and the transitional basalts of Iceland evolve towards pseudo-peritectic relations, crystallizing olivine, plagioclase, and augite, and there, dependent on the amount of crystal fractionation, will either be fully consumed or will cross peritectic relations. The fact that low-Ca pyroxene is not observed in any of the Icelandic Fe-Ti basalts suggest that these do not reach invariant relations. The Galapagos basalts, having quartz normative parents, can evolve past pseudo-peritectic relations crystallizing pigeonite, augite, and plagioclase. This explains why enriched tholeiites and mildly alkalic lavas show relatively high degrees of crystallization and indicate low temperatures without reaching pseudo-invariant relations while, on the other hand, depleted tholeiites reach the invariant and pigeonite saturation at relatively smaller degrees of crystallization and significantly higher temperatures.

The crystallization of Fe-Ti oxide minerals are, for the range of basalts considered here, largely a function of the liquid composition, in particular TiO_2 and FeO [Thy, 1983b]. Therefore, Fe-Ti oxide minerals will for the mildly alkalic and transitional basalts appear before low-Ca pyroxene pseudo-invariant relations has been reached. For the depleted tholeiites, Fe-Ti oxides will appear after crystallization of pigeonite and Fe and Ti enrichments can for this reason continue at and past invariant relations. The main effect of magnetite crystallization on the liquid lines of descent is a marked silica enrichment [Mahood and Baker, 1986] and the production of andesitic liquids as seen for both the Hekla and Galapagos lavas largely "independent" on the silicate phase assemblages.

It is evident from the good correspondence between the one atmosphere, anhydrous experiments (Table 2) and the natural lavas (Table 1) that the chemical variation observed within each subgroup of lavas reflects upper crustal processes and fundamental phase equilibria constraints. In contrast, the overall range from depleted tholeiite to mildly alkalic compositions must, as generally accepted, reflect differences in deep-seated lower crustal or mantle differentiation and melting.

Schilling et al. [1982b] and Meyer et al. [1985] suggested that the variation along the Icelandic propagating rift could "be related to a mantle plume (or blob) that ascended beneath

TABLE 2. Experimental, one-atmosphere Fe-Ti basalts coexisting with augite, plagioclase, and olivine (OFZ, KFZ, 43°N, SURTSEY, PANTELL) or augite, plagioclase, olivine, and pigeonite (FAMOUS, 23°N)

	OFZ	KFZ	43N	SURTSEY	PANTELL	FAMOUS	23N
SiO_2	50.42	52.00	50.90	47.54	45.20	51.10	51.20
TiO_2	4.14	2.77	2.35	4.57	5.05	2.43	2.66
Al_2O_3	11.38	13.20	12.80	13.20	13.40	12.90	12.60
FeO	16.04	12.90	13.10	13.59	12.50	11.50	12.80
MnO	0.28	0.23	0.19	0.23	0.14	0.13	0.21
MgO	3.33	5.35	6.23	4.80	4.21	7.30	6.16
CaO	7.71	9.71	10.40	9.16	9.29	11.40	10.00
Na_2O	2.43	3.38	1.80	3.63	3.99	1.47	2.12
K2O	1.07	0.36	0.52	1.16	2.10	0.23	0.34
P_2O_5	0.46	-	-	0.61	1.61	-	-
Total	97.26	99.90	98.29	98.49	97.49	98.46	98.09
$Mg/(Mg+Fe^{2+})$	0.30	0.46	0.50	0.42	0.41	0.57	0.50
q/(q+pl+ol)	0.55	0.33	0.46	0.03	-0.11	0.48	0.47

OFZ - Enriched tholeiite from the Oceanographer fracture zone [Sample P22; Walker et al., 1979].
KFZ - Enriched tholeiites from the Kane fracture zone [Tormey et al., 1985].
43N - Depleted tholeiite from 43°N on the mid-Atlantic ridge [Grove and Bryan 1983].
SURTSEY - Mildly alkalic lava from the Surtsey volcano, Iceland [Thy, 1988].
PANTELL - Mildly alkalic lava from Pantelleria, Strait of Sicily [Mahood and Baker, 1986].
FAMOUS - Depleted tholeiite from Mt. Pluto, the FAMOUS area [Grove and Bryan, 1983].
23N - Depleted tholeiite from the mid-Atlantic ridge at 23°N [Grove and Bryan 1983].
$Mg/(Mg+Fe^{2+})$ is calculated with $Fe^{2+}/(Fe^{2+}+Fe^{3+})$=0.86.
q/(q+pl+ol) is a normative ratio, with nepheline as negative q, and based on a molecular CIPW norm with $Fe^{2+}/(Fe^{2+}+Fe^{3+})$=0.86.

central Iceland 2-3 m.y. ago and spread out laterally and triggered the southward propagation rift." The Galapagos propagating rifts is off-centered relative to the hotspot located beneath the Galapagos Islands. This suggests that the differences in the major element chemistry may be a function of the proximity to and the influence of mantle upwelling. The chemical and petrological imprint of mantle upwelling has been explored for the north Atlantic Ocean, including Iceland, by Schilling and co-workers. Of particular interest, for the present discussion, is that melting of the upwelling primordial mantle, relatively to normal ridge segments, occur over wider depth ranges and that the low SiO_2 basalts of Iceland, therefore, can be an effect of relatively high pressure melting [Schilling et al., 1983]. Meyer et al. [1985] suggested that the transition, going south from central Iceland, from tholeiitic, to transitional, and mildly alkalic basalts reflects increasing depth of melting. This model appears to be consistent with geophysical data for the region [Gebrande et al., 1980]. It, therefore, is implied that the mildly alkalic and transitional nature of the flank zone basalts of Iceland is a consequence of upwelling of hot primordial

mantle and its interaction with a normal ridge segment [Schilling, 1973; Schilling et al., 1982b].

The Galapagos spreading center basalts show a plume influenced central zone located in the proximity to the Galapagos hotspot [Schilling et al. 1982a]. Fisk et al. [1982] showed that glasses from this zone were relatively displaced away from quartz on the normative projections and were transitional and mildly alkalic. This is in contrast to the majority of glasses dredged from the propagating rifts which show relatively higher normative quartz and are depleted tholeiites [Christie and Sinton, 1986]. Fisk et al. [1982] suggested that the displacement for the central zone glasses reflected source differences. Schilling et al. [1982a] discussed the possibility that the central zone lavas could reflect higher pressure melting.

The plume model of Schilling [1973] suggests that major processes controlling the origin of the transitional and mildly alkalic lavas of spreading centers can be related to variable depth melting and compositional characteristics of the primordial mantles. Differentiation and crystallization in high-level magma chambers plays a second-order effect in determining the variability of the erupted magmas along the south-eastern propagating rift zone of Iceland.

Acknowledgments. Critical comments from M.R. Fisk, S.P. Jakobsson, and D. Walker were very helpful. The author acknowledges a National Research Council-NASA Research Associateship.

References

Baldridge, W.S., McGetchin, T.R., and Frey, F.A., Magmatic evolution of Hekla, Iceland, *Contrib. Mineral. Petrol.*, *42*, 245-258, 1973.

Christie, D.M., and Sinton, J.M., Evolution of abyssal lavas along propagating segments of the Galapagos spreading center, *Earth Planet. Sci. Lett.*, *56*, 321-335, 1981.

Christie, D.M., and Sinton, J.M., Major element constraints on melting, differentiation and mixing of magmas from the Galapagos 95.5° W propagating rift system, *Contrib. Mineral. Petrol.*, *94*, 274-288, 1986.

Clague, D.A., Frey, F.A., Thompson, G., and Rindge, S., Minor and trace element geochemistry of volcanic rocks dredged from the Galapagos spreading center: Role of crystal fractionation and mantle heterogeneity, *J. Geophys. Res.*, *86*, 9469-9482, 1981.

Clocchiatti, R., and Massare, D., Experimental crystal growth in glass inclusions: The possibilities and limitations of the method, *Contrib. Mineral. Petrol.*, *89*, 193-204, 1985.

Fisk, M.R., Bence, A.E., and Schilling, J.-G., Major element chemistry of Galapagos rift zone magmas and their phenocrysts, *Earth Planet. Sci. Lett.*, *61*, 171-189, 1982.

Fornari, D.J., Perfit, M.R., Malahoff, A., and Embley, R., Geochemical studies of abyssal lavas recovered by DSRV Alvin from eastern Galapagos rift, Inca transform, and Ecuador rift. 1. Major element variations in natural glasses and spatial distribution of lavas, *J. Geophys. Res.*, *88*, 10519-10529, 1983.

Gebrande, H, Miller, H., and Einarsson, P., Seismic structure of Iceland along RRISP-profile I, *J. Geophys.*, *47*, 239-249, 1980.

Grove, T.L., and Baker, M.B., Phase equilibrium controls on the tholeiitic versus calc-alkaline differentiation trends, *J. Geophys. Res.*, *89*, 3253-3274, 1984.

Grove, T.L., and Bryan, W.B., Fractionation of pyroxene-phyric MORB at low pressure: An experimental study, *Contrib. Mineral. Petrol.*, *84*, 293-309, 1983.

Imsland, P., Iceland and the ocean floor. Comparison of chemical characteristics of magmatic rocks and some volcanic features, *Contrib. Mineral. Petrol.*, *83*, 31-37, 1983.

Jakobsson, S.P., Chemistry and distribution pattern of Recent basaltic rocks in Iceland, *Lithos*, *5*, 365-386, 1972.

Jakobsson, S.P., Outline of the petrology of Iceland, *Jökull*, *29*, 57-73, 1980.

Jakobsson, S.P., Petrology of Recent basalts of the eastern volcanic zone, Iceland, *Acta Naturalia Islandica*, *26*, 103 pp., 1979.

Johannesson, H., Jardlagaskipan and throun rekbelta a Vesturlandi, *Natturufraedingurinn*, *50*, 13-31 (in Icelandic with English summary), 1980.

Mahood, G.A., and Baker, D.R., Experimental constraints on depths of fractionation of mildly alkalic basalts and associated felsic rocks: Pantelleria, Strait of Sicily, *Contrib. Mineral. Petrol.*, *93*, 251-264, 1986.

Melson, W.G., Byerly, G.R., Nelen, J.A., O'Hearn, T., Wright, T.L., and Vallier, T., A catalog of the major element chemistry of abyssal volcanic glasses, *Smithsonian Contrib. Earth Sci.*, *19*, 31-60, 1977.

Melson, W.G., and O'Hearn, T., "Zero-age" variations in the composition of abyssal volcanic rocks along the axial zone of the mid-Atlantic ridge, in *The Geology of North America, vol. M, The Western North Atlantic Region*, edited by P.R. Vogt, and B.E. Tucholke, Geological Society of America, 117-136, 1986.

Meyer, P.S., Sigurdsson, H., and Schilling, J.-G., Petrology and geochemical variation along Iceland's neovolcanic zones, *J. Geophys. Res.*, *90*, 10043-10072, 1985.

Oskarsson, N., Sigvaldason, G.E., and Steinthorsson, S., A dynamic model of rift zone petrogenesis and the regional petrology of Iceland, *J. Petrol.*, *23*, 28-74, 1982.

Perfit, M.R., and Fornari, D.J., Geochemical studies of abyssal lavas recovered by DSRV Alvin from eastern Galapagos rift, Inca transform, and Ecuador rift. 2. Phase chemistry and crystallization history, *J. Geophys. Res.*, *88*, 10530-10550, 1983.

Presnall, D.C., Dixon, J.R., O'Donnell, T.H., and Dixon, S.A., Generation of mid-ocean ridge tholeiites, *J. Petrol.*, *20*, 3-35, 1979.

Presnall, D.C., Dixon, S.A., Dixon, J.R., O'Donnell, T.H., Brenner, N.L., Schrock, R.L., and Dycus, D.W., Liquidus phase relations on the join diopside-forsterite-anorthite from 1 atm to 20 kbar: Their bearing on the generation and crystallization of basaltic magma, *Contrib. Mineral. Petrol.*, *66*, 203-220, 1978.

Sack, R.O., Walker, D., and Carmichael, I.S.E., Experimental petrology of alkalic lavas: Constraints on cotectics of multiple saturation in natural basic liquids, *Contrib. Mineral. Petrol.*, *96*, 1-23, 1987.

Saemundsson, K., Outline of the geology of Iceland, *Jökull*, *29*, 7-28, 1980.

Schilling, J.-G., Iceland mantle plume: Geochemical study of Reykjanes ridge, *Nature*, *242*, 565-571, 1973.

Schilling, J.-G., Kingsley, R.H., and Devine, J.D., Galapagos hotspot-spreading center system. 1. Spatial petrological and geochemical variation (83°W-101°W), *J. Geophys. Res.*, *87*, 5593-5610, 1982a.

Schilling, J.-G., Meyer, P.S., and Kingsley, R.H., Evolution of the Iceland hotspot, *Nature*, *296*, 313-320, 1982b.

Schilling, J.-G., Zajac, M., Evans, R., Johnston, T, White, W., Devine, J.D., and Kingsley, R., Petrological and geochemical variations along the mid-Atlantic ridge from 29°N to 73°N, *Am. J. Sci.*, *283*, 510-586, 1983.

Sigurdsson, H., Schilling, J.-G., and Meyer, P.S., Skagi and Langjökull volcanic zones in Iceland. 1. Petrology and structure, *J. Geophys. Res.*, *83*, 3971-3982, 1978.

Sigvaldason, G.E., The petrology of Hekla and origin of silicic rocks in Iceland, *The Eruption of Hekla 1947-1948*, part 4, vol. 1, Soc. Sci. Islandica, 44 pp., 1974.

Steinthorsson, S., Oskarsson, N., and Sigvaldason, G.E., Origin of alkalic basalts in Iceland: A plate tectonic model, *J. Geophys. Res.*, *90*, 10027-10042, 1985.

Takahashi, E., and Kushiro, I., Melting of a dry peridotite at high pressures and basalt magma genesis, *Am. Mineral.*, *68*, 859-879, 1983.

Thy, P., Phase relations in transitional and alkali basaltic glasses from Iceland, *Contrib. Mineral. Petrol.*, *82*, 232-251, 1983a.

Thy, P., Spinel minerals in transitional and alkali basaltic glasses from Iceland, *Contrib. Mineral. Petrol.*, *83*, 141-149, 1983b.

Thy, P., High and low pressure phase equilibria of a mildly alkalic lava from the 1965 Surtsey eruption: Part 1. Experimental results, Manuscript submitted, 1988.

Tormey, D.R., Grove, T.L., and Bryan, W.B., Experimental petrology of normal MORB near the Kane fracture zone: 22°-25° N, mid-Atlantic ridge, *Contrib. Mineral. Petrol.*, *96*, 121-139, 1987.

Vogt, P.R., The Iceland mantle plume: Status of the hypothesis after a decade of new work, in *Structure and Development of the Greenland-Scotland Ridge*, edited by M.H.P. Bott et al., 191-216, Plenum, New York, 1983.

Walker, D., Shibata, T., and DeLong, S.E., Abyssal tholeiites from the Oceanographer fracture zone. II. Phase equilibria and mixing, *Contrib. Mineral. Petrol.*, *70*, 111-125, 1979.

Yoder, H.S., and Tilley, C.E., Origin of basaltic magmas: An experimental study of natural and synthetic rock systems, *J. Petrol.*, *3*, 342-532, 1962.

GEOLOGY OF AXIAL SEAMOUNT, JUAN DE FUCA SPREADING CENTER, NORTHEASTERN PACIFIC

Zonenshain L.P.*, Kuzmin M.I.**, Bogdanov Yu.A.*, Lisitsin A.P.* and Podrazhansky A.M.*,

*Institute of Oceanology, the USSR Academy of Sciences, Moscow, 117218, USSR
**Institute of Geochemistry, the Siberian Branch of the USSR Academy of Sciences, 664033, Irkutsk, USSR

Abstract. Axial Seamount was studied in detail with two *Pisces* submersibles during the 12th cruise of the Soviet R/V *Akademik Mstislav Keldysh* in fall of 1986. The data reveal a three-stage evolution of Axial Seamount. In the first stage (50,000–60,000 years ago), the axial volcano was constructed from tube lava flows. The second stage began nearly 5,000 years ago with intensive sheet flow eruptions on the axial volcano summit and finished with a catastrophic event, when the summit collapsed and the central caldera appeared. During the third, or postcaldera, stage the caldera floor was flooded with sheet flow lavas that were associated with collapsed lava pits, lava lakes and superimposed high-temperature hydrothermal activity.

Introduction

The Juan de Fuca Ridge in the NE Pacific Ocean (Fig. 1) is a spreading center oriented N24°E, along which the Pacific and Juan de Fuca plates move apart at the rate of 5.9 cm/year (Wilson et al., 1984). The Juan de Fuca Ridge crest is elevated almost 1000 m above the mean mid-ocean ridge level, being 2000–2500 m below sea level. The shallower ridge depth can be explained by either closeness of the spreading center to, or coincidence with, a hot spot. Several seamount chains that extend northwest from the ridge (e.g., the Pratt-Welker, Eickelberg, Cobb seamounts) are believed to originate from the hot spot activity. The Juan de Fuca spreading center came into existence 18 m.y. ago and was inherited from the previous Pacific/Farallon plate boundary. This spreading center has been modified by a series of propagating rifts (Delaney et al., 1981; Wilson et al., 1984).

Hydrothermal activity with massive sulfide ore deposition was found recently on the Juan de Fuca Ridge (CASM RESEARCH GROUP, 1983; Crane et al., 1985; ASHES Expedition, 1986). Axial Seamount is the highest point on the Juan de Fuca ridge, reaching depths of 1500–1600 m. It has been studied with the submersibles *Pisces* and *Alvin* by the Canadian and American teams (Canadian-American Seamount Expedition, 1985; U.S. Geological Survey, 1986). Previous studies have revealed that Axial Seamount is topped by a central caldera opened southwards and closed to the north. Hydrothermal vents with massive sulfides have been found within the caldera. The caldera floor is flooded by young sheet lava flows that conceal older ore deposits. However, the geological evolution of the Juan de Fuca Ridge near Axial Seamount remains poorly known.

In this paper, we present results of the Soviet study of Axial Seamount which was carried out in September 1986 from R/V *Akademik Mstis-*

lav Keldysh with two submersibles, *Pisces* VII and *Pisces* XI, on board. The work included echo-sounding, seismic reflection study, magnetic measurements, piston coring, water sampling with the "Rosett" system, and photo and video recording from deep-tow vehicles Zvuk 4 and Zvuk-Geo. Navigation used the Loran-C system. All geological features of Axial Seamount, including the floor, walls and shoulders of the Central Caldera, were studied in detail during 23 dives with the *Pisces* submersibles.

General Description

Axial Seamount has several specific characteristics. First, its depth of 1400–1600 m is elevated nearly 500 m above the Juan de Fuca Ridge. Second, its center is occupied by a horse shoe-shaped caldera instead of an axial graben typical of fast spreading centers. Third, the mountain itself and caldera walls are oriented 150° NNW-SSE, which is oblique by 46° with respect to the spreading direction. Consequently, a significant dextral strike-slip component must occur along structural lines associated with the caldera and caldera walls. Fourth, Axial Seamount is the youngest seamount of the Cobb seamount chain which includes the prominent Cobb Seamount, Brown Bear, Grizzly and some other seamounts. This chain is believed to originate from the Cobb hot spot, which until recently was in the off-ridge position. Axial Seamount, however, is located near or at the spreading center and therefore its youngest features are spreading generated.

The rate of the Pacific plate motion with respect to the Cobb hot spot was determined by Wilson et al. (1984) to be 3.6 cm/year and by Karsten and Delaney (1986) to be nearly 5 cm/year. As the spreading rate of the Pacific plate relative to the Juan de Fuca spreading ridge is 2.9 cm/year, the ridge converges with the hot spot at a rate of 0.7–2.0 cm/year. Between 3.5–5 Ma, when the Cobb Seamount was formed, the spreading center was 25–35 km eastwards from the hot spot. The ridge axis coincided with the hot spot about 100,000–200,000 years ago. This coincidence of the spreading center with the hot spot is apparently responsible for the anomalous features of Axial Seamount. This is consistent with proposals by Delaney et al. (1981) that the Juan de Fuca Ridge recently jumped 20 km westwards to the Cobb hot spot, and Axial Seamount was formed.

Morphology

Axial Seamount has a more or less isometric shape that is slightly elongated NNW-SSE (Fig. 2). It is 10–15 km across and has a cone-like profile in cross-section, with a top cut by the central caldera (Fig. 3). The caldera depression runs NNW-SSE and is 7 km long and 2.5 km wide. It has a horse-shoe shape open to the south. The caldera walls are elevated 100–150 m above the floor, but a broad pass exists in the

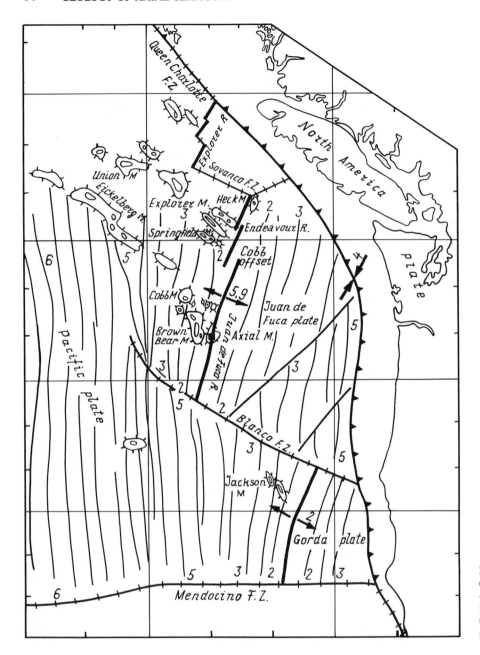

Fig. 1. Regional setting of Axial Seamount (solid square) on the NE Pacific. Thick line – spreading center, line with dashes – transform fault, toothed line – subduction zone, thin line – magnetic lineation. Propagating rift pseudofaults are also shown.

northern wall where the wall elevation is no more than 50 m. The pass is flooded by young sheet lava flows now broken by numerous open fissures, or gjar.

Geology

The Summit

The gentle slopes of Axial Seamount, 2° to 4°, coincide with the inclinations of lava flows. The summit of the ridge rimming the caldera is mostly covered by loose pelagic sediments 60–70 cm thick. The sediment thickness was determined from in situ observations during sub-

mersible dives on the northwestern caldera wall and also from piston coring on the western slope of the seamount. Pillowed tube lava flows can be seen in many places concealed under the sedimentary veneer. In places, the summit is topped by low (3–5 m high) mounds consisting of pillow lava flows.

Observations from the *Pisces* submersibles show that the Axial Seamount flanks are composed of at least three complexes of different age. The lower and oldest complex is represented by pillow lavas. They form a single volcanic pile and display patterns that indicate they were poured out from one or several centers near the Axial Seamount summit. Pillow lavas are very well exposed on the vertical caldera walls (Fig. 4a). They have a radial jointing in cross-section. Aphanitic, dolerite-like lavas with rare plagioclase phenocrysts predominate. Massive sheet flows

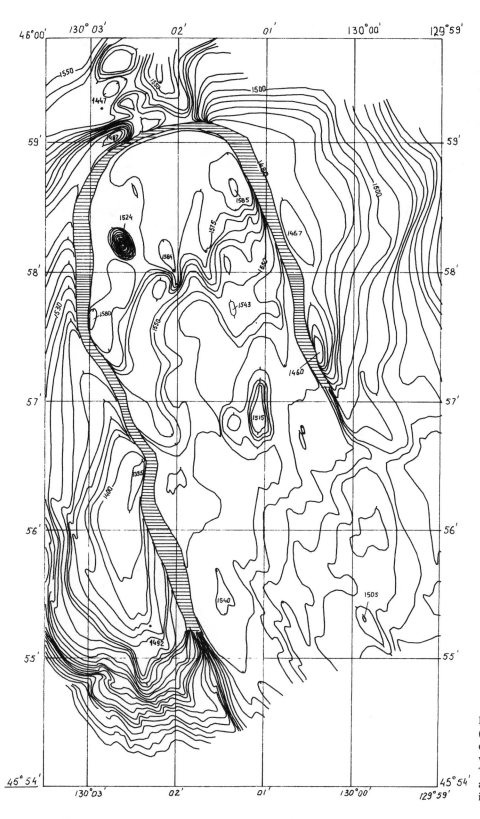

Fig. 2. Bathymetric map of Axial Seamount (based on 12 and 20 KHz narrow beam echosounding). Contours are in 5 m intervals. Horizontal striation – caldera walls. Two dots numbered correspondingly 1447 and 1492 indicate positions of cores shown in Figure 5.

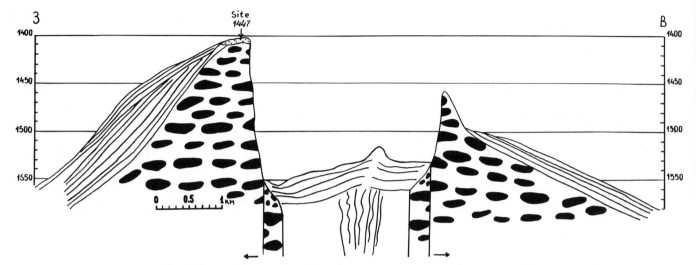

Fig. 3. Cross section through Axial Seamount. Black spots are pillow lavas older than 50,000 years, thin lines on slopes correspond to pre-caldera sheet flows 9,000–5,000 years old. The caldera floor is occupied by sheet flows younger than 5,000 years.

or sills are included among the lava tube pile. Hollow pillows were rarely observed on the summit near volcanic centers. Vertical dike-like bodies were observed during one dive along the foot of the western wall. Available data show that Axial Seamount was a central type volcano built from outpourings of tube lava flows before the roof collapse and caldera origination.

The age of pillow lavas can be estimated from the 60–70 cm thickness of overlying oozes. Bogdanov (personal communication) used paleontological and ^{14}C determinations of piston cored sediments from the top and slopes of Axial Seamount to estimate sedimentation rate at 1–2 cm in 1000 years. Accordingly the age of basalts is 60,000–30,000 years.

Loose sediments form the second complex. They fill all depressions between lava tubes and often completely overlie the lavas concealing all irregularities of lava flow surfaces (Fig. 4b). The sedimentary veneer is stratified. Two 15 cm-thick, columnar-sections were obtained by piston coring from submersibles on the north and west summits of the rimming ridge. Sediments are silt-pelitic ooze with diatoms, forams, nannofossils and include large amounts of glass. They have a smooth upper surface with no traces of bioturbation by bottom organisms. Hyaloclastics are distributed unevenly along the columnar section and are concentrated at several levels. In both cores there are two levels mostly enriched with basalt glass clastics (Fig. 5); they are in the upper 1–2 cm and between 4 and 12 cm from the top. Three or four independent thin layers of glass enrichment can be recognized within the second (lower) level. Enrichment with basaltic glass clastics is interpreted to indicate volcanic episodes. The stratification shows that volcanism was discontinuous: intervals without eruptions gave way to violent volcanic activities. Saidova and Mukhina (unpublished results) used planktonic and benthic foraminifera distributional patterns within the cores to correlate the sections with the Holocene paleoclimatic scale. They revealed that the cold Subatlantic, Subboreal and Boreal periods are characterized in the cores by small amount of forams (less than 300 per 10 g of sediments), whereas the warm Atlantic period is marked by increasing amounts of forams up to several thousands per 10 g of sediments. This pattern may be related to either decrease or increase of influence of the high productivity California current. Saidova and Mukhina found that a Subboreal community 4,500 years old is 4 cm below the core top and the Atlantic community 7,500 years old is 8 cm below the core top. The

whole age of the piston core sections is about 12,000 years . The lower layer enriched with hyaloclastics belongs to the warm Atlantic period (9,000 to 5,000 years ago). It can be inferred that the non-volcanic interval lasted from 12,000 to 9,000 years, then intensive volcanic activity took place from 9,000 to 5,000 years and embraced 3–4 independent episodes. No volcanism occurred from 5,000 to 3,000 years ago.

The third complex mostly consists of sheet lava flows, which overlie pillow lavas of the first complex. The contact between the two units is very well exposed in the west wall of the caldera. The thickness of sheet flows in this outcrop is no more than 20 m. Sheet flows appear to cover all the slopes of the Axial Seamount. They were erupted as thin, 8- to 10-cm-thick layers covering a large area. Abundant glassy crust debris and glass clastics are typical for these sheet flows. Their upper surface often has a ropy structure and collapsed lava pits are common. The sedimentary veneer is very thin, no more than 3–5 cm thick. Basalts of the sheet lava unit probably supplied sediments overlying the pillow unit with glass clastics, and are thus apparently correlated with the 5000–9000 year old layers enriched with hyaloclastics.

The pass in the north caldera wall is 80 m lower than the rimming ridge. The path is flooded by sheet flows of the third complex. Pillow lavas are preserved only at the top of small mounds elevated over the sheet flows. Numerous open fissures or gjar are characteristic features of the pass. Gjar have the orientation NW 330°-SE 150°, i.e., they follow the orientation of the caldera but run 40–50° obliquely to spreading direction. The gjar width varies from 1 to 5 m in some places; fissures widen up to 10–15 m and then narrow to a very thin, 0.5 m wide, vertical fissure cutting the lava pile. Such irregularities in the fissure width along strike can be explained by their oblique position relative to the spreading direction. Apparently, not only pure extension but also strike slip motion with a dextral component of displacement occurs there. Gjar of the north pass appear to be young features which are active now. Besides active gjar some inactive gjar filled with sediments were observed on the western slope of the mountain.

Caldera Walls

The horse-shoe shaped caldera walls are formed by vertical scarps divided by areas of angular talus blocks. The scarps are 20–50 m high and usually are parallel to one another. They are very impressive fea-

Fig. 4. Seafloor photographs on Axial Seamount, taken from *Pisces* submersibles. a. (top left) the western caldera wall, which cuts pillow lava
flows. b. (bottom left) the summit of the northern ridge covered by sediments, 50–60 cm thick. c. (top right) large blocky talus at the foot of
the caldera wall. d. (middle right) lobate lava flows on the top of a
"frozen" lava lake on the caldera floor. e. (bottom right) collapsed lava
pit on the caldera floor.

tures when observed from submersibles. Pillow lavas are cut as if by a knife, and commonly protrude as a cornice hanging over the precipice below. The nature of the scarps strongly suggests they were formed by collapse which occurred by gravitational downfall of blocks, that is without listric faulting or other types of sliding of blocks. The pattern here is completely different from slow spreading ridges with their tilted and faulted terrain. Large blocky talus descends like a cone downslope. It consists of blocks in size from 10–20 cm to 1–5 cm (Fig. 4c). Many talus accumulations are active now, and blocks begin to move downslope each time the submersible touchs them even slightly. The main volume of the talus accumulations appears to have formed simultaneously with the caldera development, as blocks are covered by sediments several centimeters thick. But the main evidence is that the talus blocks are overlain in many places by caldera floor lavas. Lavas overlying talus blocks can be seen in almost any place at the foot of the wall. Caldera floor lavas filled free spaces between the blocks which are buried under younger flows. Sometimes small pillow lava fragments were captured by caldera lava flows and traveled with them some distance from the wall. Some blocks are covered by pale and orange hydrothermal deposits, which may be bacterial mats.

Caldera Floor

The caldera floor is 2.5–2.8 km wide. It deepens to the north where the 1580–1590 contour delineates the main caldera deep (Fig. 2). To the

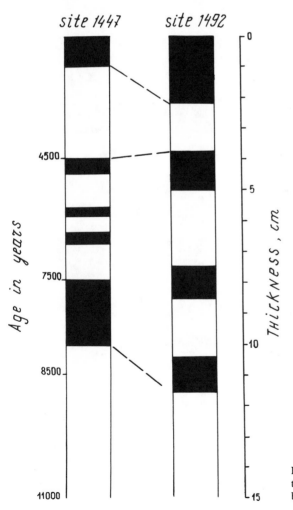

south, the floor shoals gradually to 1550–1520 m, where it is occupied by irregular, gentle volcanic mounds 10–20 m high. A hill, 100 m across, with steep, almost vertical slopes elevated 60 m high over the caldera floor, occurs in the NW part of the caldera. It is composed of pillow lavas that are overlain by sheet flows with abundant plagioclase phenocrysts and by a thin sedimentary cover. According to paleontological evidence (Saidova and Mukhina, unpublished results), sediments near the bottom belong to the Atlantic period, i.e. they are 5000–9000 years old, being much older than the very recent caldera floor basalt. This hill may be some kind of kipuka as described by Kappel and Normark (1987) on the southern Juan de Fuca Ridge, and appears to be a remnant of the collapsed caldera roof.

The caldera floor is completely covered by sheet lava flows. Several types of lavas can be recognized (Fig. 6). Lobate lavas (Fig. 4d) are the most common type. They extend over hundreds of meters and indicate calm lava outpouring with a low extrusion rate and have not traveled too far from a temporary lava chamber. Bulbous pillows (1 m in diameter) with closely spaced cracks commonly protrude over the lobate lava surface. These lava bulbs and contracting cracks mark the sites of short lava outbursts through the lobate surface of the lava chamber roof.

Numerous gaps or collapsed lava pits (Fig. 4e) exist on the lobate surface, revealing large hollows under the hard lava surface. The upper crust is completely collapsed in the central parts of previous lava chambers and many hollow lava pillars (Fig. 7a,b) are standing up from the floor encircling a central place where pillars are lacking. The pillars have all the features described earlier on the East Pacific Rise (Francheteau et al., 1979). Many horizontal rims and ledges on the pillars mark gradual subsidence of the level in the lava lake as lava was removed from the temporary lava chamber. The question is, to where is the lava gone? The caldera floor is more or less even and of uniform age throughout. That means that lava did not flow laterally. In part, lava level changes may result from escaping water, which was buried under lavas during the fast flooding. We infer that the main cause for subsidence in the lava level has been a decrease of the lava head from beneath. Surficial lava chambers are apparently connected by more or less wide feeding channels with the subcrustal magma chamber. The lava level oscillates in the feeding channels and correspondingly in surficial lava chambers. Lava upwelling supplies lava flows, lava downwelling produces hollows under the solid upper lava crust and collapsed lava pits. Collapsed lava pits corresponding to drained back lava lakes are numerous along the foot of the western wall. If these collapse pits were to coalesce, they could form a chain, which resembles the axial cleft in the southern Juan de Fuca Ridge (Kappel and Normark, 1987). Thus, the lobate surface of lava flows is (or was) a roof for lava columns rising from the magma chamber to the sea floor.

Ropy lavas occur as another sheet flow type. They form flat flows covering relatively large areas (Fig. 7c,e). The ropy surface is a characteristic feature of these lava flows. We called such flat surfaces "airfields". Lava whorls occur more or less commonly among ropy flows. Ropy lava fields are adjacent to lobate flows and, so, are aside from lava upwelling sites. They mark the areas where fluid lavas flowed rapidly and freely, without any obstacle. Ropy lavas consist of very thin, usually 8–10 cm thick, individual lava layers divided by glassy crust lenticules.

Another type of sheet flow is hummocky lavas, formed by strongly contorted lava sheets with a great amount of volcanic glass. Locally lava flows were observed to blend into small hollow tubes. Numerous elongated hollows, 5–10 cm long, are peculiar to this lava type. They appear to be remnants of sea water captured during fast eruptions which afterwards escaped. Lavas are piled up in wild disorder, protruded as brush-

Fig. 5. Columnar sections for two stations obtained by coring from submersibles. Station positions are given in Figure 2. Shaded areas correspond to layers enriched with basaltic glass clastics, blank – silt-pelitic oozes.

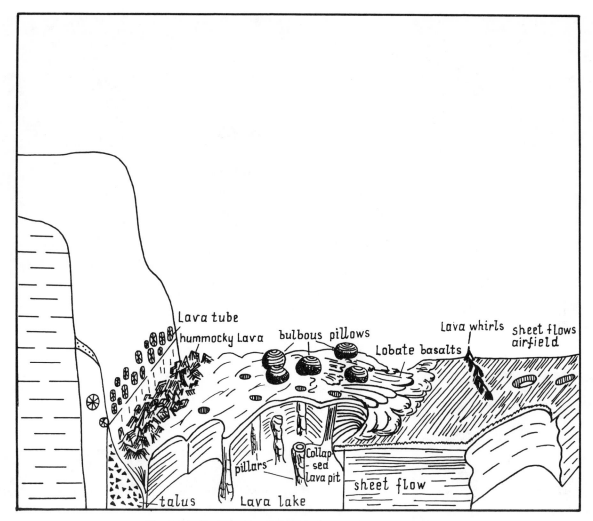

Fig. 6. Distribution of the different lava types within the caldera floor
of Axial Seamount.

es, or gathered in heaps (Fig. 7d). Hummocky lavas usually appear where calm flowing lavas meet an obstacle and collide with it. They resemble a'a lavas of terrestrial basaltic eruptions. Their chaotic piling up can be easily explained by instantaneous breaking of the upper solid crust and outburst and "pulverization" of new portions of fluid lavas which momentarily solidified as volcanic glass. Hummocky lavas were observed near the caldera walls, along boundaries of two different age lava flows and along the boundary of two simultaneous flows moving towards one another. Hummocky lavas supply near-bottom sediments with great amounts of hyaloclastic material.

All the morphological lava types can be formed simultaneously during the same volcanic episode and from the same volcanic center. The center itself can be imagined as a large lava lake connected by a wide feeding channel to a subjacent magma chamber. According to seismic reflection records from the southern Juan de Fuca Ridge (Morton et al., 1987), the magma chamber top may be no deeper than 2.5 km under the sea floor. This figure is an estimate for the length of a vertical lava column connecting the shallow chamber with surficial eruptions feeding underwater lava lakes. The distribution of the different lava types on the caldera floor is shown in Figure 6; hummocky lavas comprise

a strip 50–100 m wide along the western caldera wall, giving way to lobate lavas with collapsed pits and pillar structures towards the center of the caldera. They are succeeded by ropy lava flows.

Submersible dives revealed lava flows of two ages in the caldera floor. Following the classification of Ballard et al. (1979): the youngest lavas had brightly glittering glassy crusts almost lacking sediments; slightly older ones are covered with a 0.5–1.0 cm thick sedimentary veneer.

Volcanic processes almost completely have masked evidence of tectonic activity in the caldera. Open fissures, or gjar, so typical for spreading zones, are very rare on the caldera floor. They were mapped only in the north of the caldera near the bordering wall, along-strike of the gjar of the north pass. Fissures are oriented in the same NNW-SSE direction as those that cut the caldera rim. These fissures have localized hydrothermal activities including massive sulfides, bacterial mats and vestimentifera bushes. One of the biggest fissures near the northern wall has a zigzag shape widening along strike up to 10 m and narrowing along NW-SE segments to 2–3 m. This can be related to a dextral strike slip component in motion along the fissure, a result of the oblique position of the Axial Seamount with respect to the spreading direction. Some gjar can be explained by their immediate use as channels for magma

a   b   c

Fig. 7. Seafloor photographs on Axial Seamount, taken from Pisces submersible. a and b. (left) hollow lava pillars under the lobate lava roof. c. (top right) flat sheet flows with ropy surfaces ("airfield"). d. (middle right) hummocky lavas near the caldera wall. e. (bottom) the margin of the sheet flow domain.

d   e

upwelling (Canadian-American Seamount Expedition, 1985). The geology of the caldera is summarized in Figure 8. Submersible tracks are shown in Figure 9.

Petrology of Basalts

Basalts of Axial Seamount belong to three age groups: (1) older than 60,000–30,000 years, (2) 9,000–5,000 years old and (3) younger than 3,000 years. Morphologically, they include two groups: (1) tube lava flows or pillow basalts, which are characteristic only for the older generation and (2) sheet flows.

Pillow basalts are mostly aphyric. External parts of pillows consist of glassy basalts with very rare plagioclase (An 60) and olivine phenocrysts. Interiors have doleritic textures composed of lath-shaped plagioclase crystals and olivine microlites.

Sheet lava flows are covered by glassy crusts up to 5 cm thick, which served as thermal insulators, under which lava became well-crystallized. Phenocrysts, most commonly plagioclase and, rarely, olivine, are more abundant within the glassy crust. Interiors of lava flows are mostly aphyric, sometimes spheroidal, and, near the bottom of flows, doleritic.

Basaltic samples obtained from the isolated hill (a possible remnant of the collapsed roof) in the inner part of the caldera (depth mark 1524,

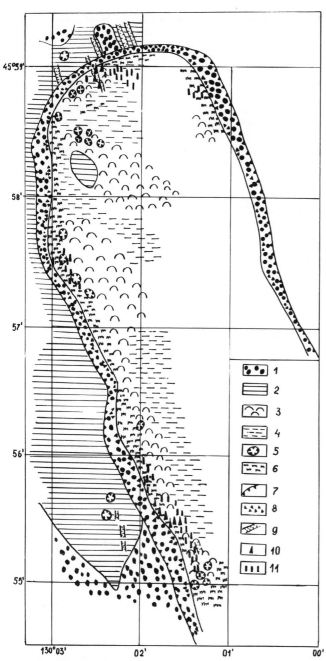

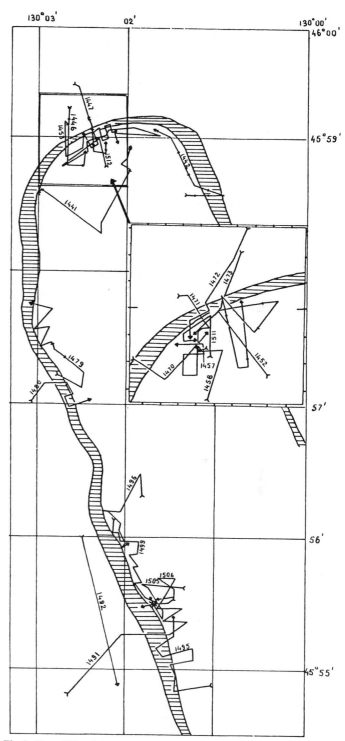

Fig. 8. Geological map of Axial Seamount, based on bathymetry, submersible observations, and near-bottom photo recording. 1. pillow lavas older than 50,000 years; 2. pre-caldera sheet flow of 9,000–5,000 years old; 3–6. caldera floor sheet flow younger than 5,000 years (3. lobate lavas, 4. ropy sheet flows or " airfields", 5. collapsed lava pits, 6. hummocky lavas); 7. upper break of the c aldera wall; 8. blocky talus; 9. open fissure; 10. hot vent; 11. ochre.

Fig. 9. *Pisces* XII and *Pisces* XI Submersible tracks on Axial Seamount. Only tracks relevant to geological data are shown.

Fig. 2) demand special consideration. A small amount of sediments was dredged together with basalts from this site. As mentioned above, these sediments are between 9,000–5,000 years old. Glassy basaltic crusts lying immediately beneath the sediments, apparently belong to the top

of the pre-caldera sheet flow unit. The basaltic glass is enriched with plagioclase phenocrysts (up to 20–25%); dimensions of individual crystals are on the average 8 × 3 mm. Besides individual phenocrysts, glomerophyric plagioclase aggregates with clinopyroxene also are present.

TABLE 1. Composition (oxides %, elements ppm) of the Juan de Fuca basalts (Axial Seamount) in the northeastern Pacific Ocean

Basalt type	SiO_2	TiO_2	Al_2O_3	Fe_2O_3	FeO	MnO	CaO	Na_2O	K_2O	nnn*	Rb	Ba	Sr	Zr	Cr	Ni	Co	V	n**
Pillow basalt	48.58	1.47	14.36	1.90	9.90	0.2	12.25	2.75	0.20	0.60	2.0	48	130	135	283	84	54	282	6
Precaldera sheet flow	48.45	1.49	14.57	1.63	10.37	0.2	12.21	2.77	0.19	0.51	2.3	43	132	137	327	101	55	272	3
Caldera floor sheet flow	48.62	1.51	14.56	1.99	9.77	0.2	12.24	2.82	0.20	0.49	1.9	52	132	152	324	100	52	302	31
Kipuka basalts	47.79	0.98	19.36	1.30	7.45	0.14	13.14	2.46	0.11	0.13	–	35	140	90	242	85	40	177	1
Axial Seamount basalts, mean values	48.61	1.50	14.53	1.95	9.82	0.2	12.25	2.81	0.20	0.51	1.9	52	132	150	320	99	53	294	42
MORB, after Wedepohl (1981)	49.14	1.17	15.64	2.64	6.66	0.16	11.84	2.40	0.20	1.09	4.9	48	134	85	317	144	45	252	

*$H_2O + H_2O$
**number of analyses

Phenocrysts have a well-defined zonal pattern, with 10–16 zones around the core. The microprobe analysis of glasses showed reversed zoning with anorthite contents up to 95–96% in the rim of the phenocrysts. These plagioclase phenocrysts settled from the melt within the magmatic chamber itself. The large amount of phenocrysts indicates long lasting subvolcanic residence, leading to plagioclase accumulation preceding eruptions. Gabbroic xenoliths with a large amount of plagioclase phenocrysts are found elsewhere in the Juan de Fuca Ridge (Dixon et al., 1986). This can be evidence for capturing previously stored crystals by new magma portions during replenishment of a former magma chamber. This crystal storage shows that a long nonvolcanic period existed when fractionation may have taken place.

Compositionally all of the basalts, including pillow lavas and sheet flows, are MORBs, being low-potassium tholeiites (Table 1). No influence of the Cobb hot spot can be recorded on the basalt composition. This is supported by $^{87}Sr/^{86}Sr$ data data equal to 0.70233–0.70267 (Eaby et al., 1984), typical for MORB tholeiites. As was found earlier (Vogt and Byerly, 1976), the Juan de Fuca basalts are enriched with iron. The Fe/(Fe+Mg) coefficient is equal to 0.61 compared with 0.52 typical for normal MORB. This high value may reflect the high degree of differentiation of the magma chamber.

Geological Evolution

The known geological history of Axial Seamount numbers no more than 60,000 years, which is a short span in the geological scale. However, during this time the Pacific plate separated from the Juan de Fuca plate no less than 3 km. The caldera width is approximately the same; consequently all the caldera floor may have been accreted due only to plate divergence. This must be taken into account when considering the geological evolution of the area. For example, we cannot expect to find remnants of the collapsed caldera roof under caldera floor sheet flows. Such remnants should be preserved only near the caldera walls. The main part of the caldera floor has to be considered as newly created oceanic crust. Evidence for separation of the caldera walls are that the caldera is open to the south and has a low pass with numerous gjar cutting its north wall.

Geological data allow us to divide the Axial Seamount history into three stages : (1) pre-caldera stage, (2) collapse and development of the caldera and (3) caldera stage. The evolution is shown schematically in Figure 10.

The central volcano composed of pillow lava flows was formed in the pre-caldera stage between 60,000–30,000 years ago. This volcano was elevated 300–400 m above the adjacent floor and was 4–5 km wide. Formation of a central volcano at the ridge axis would be similar to the behavior observed at slow-spreading ridges. However, our magnetic data show the volcano did not coincide with the magnetic zero line. It had an off-ridge origin and possibly was connected with the Cobb hot spot.

The next episode of the same pre-caldera stage was the volcanic quiescent interval that lasted several tens of thousands years (from 50,000 to 10,000 years, depending on the age determination of the pillow lava unit). The sedimentary cover on the pillow lava unit was deposited at that time.

The caldera stage began with eruptions of sheet lava flows on the flanks of Axial Seamount. The time of eruption is recorded in piston cores by thin sedimentary layers enriched with basaltic glass clasts within the interval 10–5 cm below the sea bottom. This volcanic episode occurred between 9,000 and 5,000 years ago. In contrast to the previously calm volcanic eruption of pillow lava units, volcanism was very intensive at that time. During this short time lavas covered the whole surface of the Axial Seamount, but were most abundant near the central fissure along which the axial volcano was broken nearly 5,000 years ago.

The post-caldera stage began with extensive volcanic eruptions on the caldera floor. Practically, volcanic activity did not stop at all, but continued on the caldera floor rather than on the collapsed summit. Intermittent lava lakes existed on the caldera floor during the last stage. The

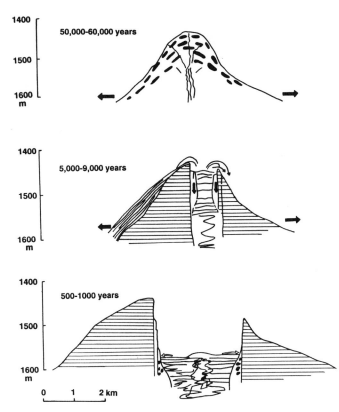

Fig. 10. Diagram showing the evolution of the Axial Seamount.

last volcanic episode was no earlier than 500 years ago. Open fissures near the north wall developed afterwards and a system of hydrothermal circulation was formed. This story is somewhat consistent with the non-steady state evolutionary model of the Juan de Fuca Ridge proposed by Kappel and Ryan (1986) and by Kappel and Normark (1987). On the southern Juan de Fuca Ridge, older pillow lavas covered by sediments also are recorded from the rims of the axial valley, and young sheet flows from the axial valley floor developed after collapse of the axial ridge summit (Kappel and Normark, 1987). Kappel and Normark (1987) assumed that a non-volcanic episode preceded the summit collapse. Data from Axial Seamount indicate that this was not the case: instead, a period of intensive volcanism began just before the summit collapsed.

Conclusions

The geological history of Axial Seamount on the Juan de Fuca Ridge shows that a spreading center was superimposed on an off-ridge volcano. The nature of the volcano is unclear: either it is hot spot-related or represents off-axis volcanism of the spreading center. Superposition of the spreading center on the previous central volcano led to a catastrophic event 5,000 years ago when the magma chamber roof collapsed and the central caldera developed. A calm period from 9,000 to 5,000 years ago preceded the catastrophe. Melt storage and crystal fractionation took place in the developing magma chamber just before the commencement of the new volcanic activity related to spreading processes. As the plate separation site shifted to Axial Seamount only recently, the spreading until now has been oblique relative to the spreading direc-

tion, and is trying to adjust by emplacement along fissures parallel to the spreading direction. Thus, we have a new-born spreading feature on Axial Seamount.

Acknowledgments. We wish to express our gratitude to the submersible team and to the Captain and crew of the R/V Academik *Mstislav Keldysh* for their support during submersible dives. We thank Jill Karsten and John Sinton from the University of Hawaii and R.L. Chase from University of British Columbia for their reviews, numerous helpful suggestions and hard work to improve English grammar and spelling.

References

ASHES Expedition, *Pisces* submersible exploration of a high-temperature vent field in the caldera of Axial Volcano, Juan de Fuca Ridge (abstract), *Eos Trans. AGU,* 67, 1027, 1986.

Ballard, R. D., R. T. Holcomb, and T. H. van Andel, The Galapagos Rift at 86°W: 3. Sheet flows, collapse pits, and lava lakes of the Rift Valley, *J. Geophys. Res.,* 84, p. 5407–5422, 1979.

Canadian-American Seamount Expedition, Hydrothermal vents on the Axial Seamount of the Juan de Fuca Ridge, *Nature,* 313, p. 212–214, 1985.

CASM RESEARCH GROUP, North caldera hydrothermal vent field, Axial Seamount, Juan de Fuca Ridge (abstract), *Eos Trans. AGU,* 64, 723, 1983.

Crane, K., F. Aikman, R. Embley, S. Hammond, A. Malahoff, and J. Lupton, The distribution of geothermal fields on the Juan de Fuca Ridge, *J. Geophys. Res.,* 90 , p. 727–744, 1985.

Delaney, J. R., H. P. Johnson, and J. L. Karsten, The Juan de Fuca—hot spot—propagating rift system, New tectonic, geochemical and magnetic data, *J. Geophys. Res.,* 86, 11,747–11,750, 1981.

Dixon, J. E., D. A. Clague, and J.-P. Eissen, Gabbroic xenoliths and host ferrobasalt from the Southern Juan de Fuca Ridge, *J. Geophys. Res.,* 91, 3795–3828, 1986.

Eaby, J., D. A. Clague, and J. R. Delaney, Sr isotopic variations along the Juan de Fuca Ridge, *J. Geophys. Res.,* 89, 7883–7890, 1984.

Francheteau, J., T. Juteau, and C. Rangin, Basaltic pillars in collapsed lava-pools on the deep ocean floor, *Nature,* 281, 209–211, 1979.

Kappel, E. S., and W. R. Normark, Morphometric variability within the axial zone of the Southern Juan de Fuca Ridge, Interpretation from Sea MARC II, Sea MARC I and deep-sea photography, *J. Geophys. Res.,* 92, 11,291–11,302, 1987.

Kappel, E. S., and W.B.F. Ryan, Volcanic episodicity and a non-steady state rift valley along northeast Pacific spreading centers: Evidence from Sea MARC I, *J. Geophys. Res.,* 91, 13,925–13,940, 1986.

Karsten, T. L., and J. R. Delaney, Hot spot/migrating ridge crest interaction—Juan de Fuca Ridge style (abstract), *Eos Trans. AGU,* 67, 1254, 1986.

Morton, J. L., N. H. Sleep, W. R. Normark, and D. H. Tompkins, Structure of the southern Juan de Fuca Ridge from seismic reflection records, *J. Geophys. Res.,* 92, 11,315–11,326, 1987.

U.S. Geological Survey Juan de Fuca Study Group, Submarine fissure eruption and hydrothermal vents on the southern Juan de Fuca Ridge: Preliminary observations from the submersible *Alvin, Geology,* 14, 823–827, 1986.

Vogt, P. R., and G. R. Byerly, Magnetic anomalies and basalt composition in the Juan de Fuca-Gorda Ridge Area, *Earth Planet. Sci. Lett.,* 33, 185–207, 1976.

Wedepohl, K. H, Tholeiitic basalts from spreading ocean ridges. The growth of the oceanic crust, *Naturwissenschaften,* 68, N3, 110–119, 1981.

Wilson, D. S., R. N. Hey, and C. E. Nishimura, Propagation as a mechanism of ridge reorientation of the Juan de Fuca Ridge, *J. Geophys. Res.,* 89, 9215–9226, 1984.

GEODETIC AND GEOPHYSICAL EVIDENCE FOR MAGMA MOVEMENT AND DYKE INJECTION DURING THE KRAFLA RIFTING EPISODE IN NORTH ICELAND

Wolfgang R. Jocoby[1], Hannsjörg Zdarsky[1] and Uta Altmann[2]

[1]Institut für Geowissenschaften, Johannes Gutenberg-Universität, Mainz, West Germany
[2]Institut für Meteorologie und Geophysik, Johann Wolfgang Goethe-Universität, Frankfurt a. M., West Germany

Abstract. An accretionary episode occurred along about 100 km of Mid-Atlantic Ridge axis in N-Iceland since 1975. Repeated observations of point positions, elevations and gravity in the region (by various Icelandic and German groups) have provided ample data on local and regional deformation and gravity change (observed Δg, Free Air anomaly ΔFA, and Bouguer anomaly ΔBA). From these data it is possible to model, at least in part, the source geometry of deep (upper mantle) magma transport, of magma chambers below the Krafla volcanic complex, and of dyke injections into the accretionary fissure swarm. The three data sets of horizontal and vertical displacements and of gravity can, however, not be fitted exclusively by the source characteristics as geometry, pressure, dislocations, and mass or density, but they constrain also the lateral and vertical elasticity structure of the confining medium; an elastic half space is too simple a model.

We have applied analytical and numerical modeling and least-squares optimization; in spite of the model simplicity the results are quite satisfactory. The increments of dyke injection agree well with the seismicity in space and time characterizing the individual injection events. The short-wavelength BA correlates with fissure swarms belonging to presently inactive volcanic systems. The long-wavelength BA indicates that hot, low-density magma moved into the region along the axis of the neovolcanic zone at crust-mantle transition depths. This transition must be modeled as a region of magma accumulation. Seismicity suggests rather long-distance hydraulic connections in the axial rift zone which is interpreted to represent the "deep" expression of the plate boundary.

1. Introduction

The Krafla rifting episode is the best observed of its kind and data on surface deformation and gravity changes enable us to quantitatively estimate magma movements within the complicated plumbing system at depth. Magma movements cause volume and mass changes in some space-time distribution. The ultimate aim would be to obtain a complete description of it, but we are limited by the non-uniqueness problem and by the limited sensitivity of the effects in some parts of the model space. We can however hope to test ideas and models and improve understanding of underlying geodynamic

processes as the relative importance of magma ascent or migration and that of plate divergence for the dynamics of rifting. We assume that there are at least similarities between normal sections of the mid-ocean rift system and the anomalous one of Iceland.

In this paper, after describing the Krafla volcanic system in the active neovolcanic zone, the rifting episode since 1975, and our conceptual model of the deep plumbing system, we describe the new data. We treat the horizontal deformation and the formation of crustal dykes, then the vertical movements and gravity changes and what they tell us about processes in the crust-mantle transition zone. Finally we speculate on how the magma gets from there to the dykes.

The geometric setting of Iceland and the Krafla volcanic system have been described in a number of papers, of which a few are quoted: Pálmason and Saemundsson [1974] review the geology and geophysics of Iceland; Björnsson [1985] and Björnsson et al. [1977, 1979] describe the recent Krafla activity and Walker [1974] describes a similar fossil one. Recent collections of papers on these subjects are in Jacoby et al. [1980] and Steinthorsson and Jacoby [1985].

The most important aspect (Fig. 1) is that the Krafla volcanic complex and its fissure swarm is one of a set of similar volcanic systems arranged en echelon along the northeastern neovolcanic zone. These volcanic complexes form a "volcanic axis" that connects the centers of Bárdabunga and Kverkfjöll at 65°N at about the center of Iceland, considered also the center of the Iceland hot spot, with the Tjörnes fracture zone near the north coast. The fissure swarms are oriented almost normal to the theoretical spreading direction as given by Minster and Jordan [1978]. The volcanic axis trends NW, about 20° oblique to the fissure swarms. The two active Husavik and Grimsey faults in the Tjörnes zone have a more northerly trend toward the offset Kolbeinsey ridge than the theoretical spreading direction and are thus not strictly transform faults.

A foreboding of the current/recent rifting episode was an increased seismicity, partly in swarms, during 1975 and earlier [Björnsson et al., 1977, 1979]. The rifting episode proper began with a small volcanic eruption at Leirhnukur near the caldera center, followed by rifting on old and new fissures and strong seismicity near the coast (submarine), about 50 km to the north. A magnitude $M_s \approx 6.5$ event also took place on the Grimsey fault near Kopasker with a typical transform strike-slip faulting mechanism [Einarsson, 1986, 1987]. As the rifting and earthquake activity proceded the caldera floor rapidly subsided in bowl shape: the underlying magma chamber at about 3 km depth [Björnsson et al., 1977] deflated. Uplift of the caldera floor then resumed, indicating re-pressurizing and inflation

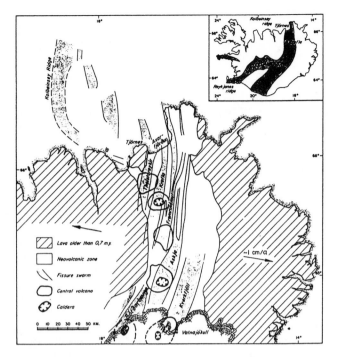

Fig. 1. Tectonic sketch map of Northeast Iceland (after Björnsson [1985]).

of the magma chamber, until after several months another deflation event took place accompanied by renewed seismicity and rifting in the fissure swarm, this time closer to the caldera.

The inflation-deflation activity continued until now with 17 major deflation events. Nine were accompanied by fissure eruptions. The latest and most voluminous eruption occurred in September 1984. Since then the caldera floor has been slowly elevated to a maximum ever with somewhat variable rates, but no deflation has occurred. The eruptions were within and outside the caldera; the magma chemistry differed abruptly at the northern rim fault, with the hotter and more magnesium-rich melts outside [Grönvold and Mäkipää, 1978].

There is ample evidence that during deflation events magma is injected into the fissures. Migrating seismicity that accompanied the events was interpreted to follow the advancing subhorizontal magma flow [Einarsson and Brandsdóttir, 1980]. In view of the abrupt chemical change of the lava at the caldera rim the above interpretation might be questioned and it might be suggested that the breaking open of the magma chamber may trigger direct flow from greater depth into the fissures without deep seismicity. Thus the path of the magma remains obscure.

To help understand the complex system and design discriminating tests, we use a speculative conceptual model of the plumbing system under central and northern Iceland as below.

Marquart and Jacoby [1985] interpreted the "magmatic axis" (see above) to be the deeper, i.e. upper mantle, expression of the plate boundary. In the neovolcanic zone of Iceland the lithosphere consists essentially only of the crust. The plate boundary is oblique to the spreading direction, and where fissure swarms which are normal to spreading intersect the deeper boundary, volcanic complexes and magma chambers develop. The magmatic axis has the highest temperatures and melt fractions and the lowest viscosities. It can

be expected to channel hydraulic pressure transmission and magma flow if there is a horizontal pressure gradient as is expected near the head of a mantle plume. It is assumed that the Bárdarbunga and Kverkfjöll volcanic centers are approximately the center of the Iceland mantle plume and that the plume convects internally on a length scale of about 100 km such that there is flow below the plates in radial direction at velocities greater than the plate divergence rate. It is likely that the axis directs a large portion of this flow [Vogt, 1976].

In the rising column below the center of the Iceland hot spot material is progressively melting. The melt collects into growing pockets which rise faster than the sluggish cristal mush. This keeps the melt fraction generally at a rather low value, say 5%, as estimated by Schmeling [1985]. The melt accumulates in the crust-mantle transition region because of the lower crustal density especially at the magmatic axis where a melt fraction of 20% has been estimated [Schmeling, 1985] from the low seismic velocities and electrical resistivities.

The manner in which melt rises is not known, but it appears likely that it is not steady in time. Occasionally a bigger "blob" arrives in the melt accumulation zone. It will then upset the pressure equilibrium and create a transient pulse migrating through the plumbing system. Perhaps such an event occurred near the plume center before and during the Krafla rifting episode. Enhanced seismicity below the Bárdarbunga volcanic complex involving several strong earthquakes with thrust mechanisms indicate the deflation of a magma reservoir or blob that emptied itself into the magma-rich zone of the crust-mantle transition at the plate boundary, and a hydraulic pressure was cascaded north into the Krafla system. This system acted as a valve for the magma that was partly injected into the yielding fissures to form new dykes and partly escaped to form flood basalts. Another valve sufficiently open may have been the Grimsvötn volcanic complex below the Vatnajökull icefield where eruptions occurred in the same time interval.

The connection between Krafla and Bárdarbunga may, however, also be seen in another way. The escape of melt through the Krafla system may have lowered the pressure in the whole magmatic axis and thus also in the Bárdarbunga magma chamber, and this pressure decrease resulted in the thrust earthquakes in the roof of the chamber. This view has been presented by P. Einarsson (pers. comm., 1986). We cannot decide which view is correct.

We consider each of the volcanic systems as a valve. The instantaneous state of each depends on the properties of the wall rocks and the recent volcano-tectonic history determining the normal stress on the walls of the fissures and the magma chamber. Whether magma pressure (or its increase) can open the valve depends on how tightly it is held together by the normal stress. Plate divergence will lessen it gradually and thus open the valve with time. The Krafla system must have been most ready to open, being relatively hot and soft from the previous event about 250 years ago, implying a $250 \times 0.02 = 5$ m separation. Askja, also a very active system, has had less time since its latest events (1875, 1920-29, 1961). Because of lack of data the above picture is, of course, highly speculative, but some of its facets may be testable with the data available.

2. Data

Our principal data set are cumulative horizontal and vertical point displacements and gravity changes observed at intervals before and during the rifting episode. The measurements were done by geodesists from Braunschweig Technical University [e.g. Möller

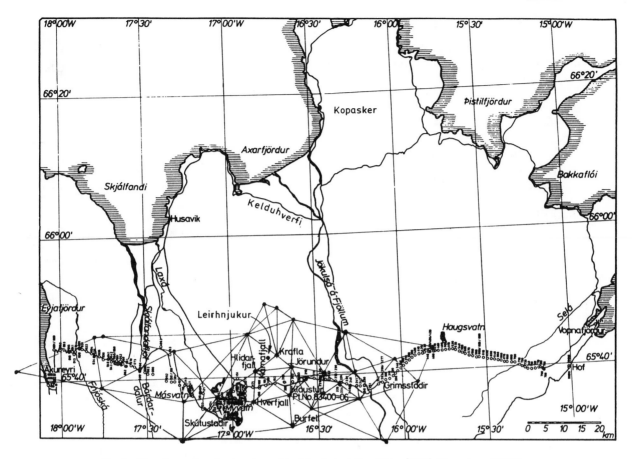

Fig. 2. The observational network used in this study (after Ritter [1982]; Kanngieser [1982]).

and Ritter, 1980; Ritter, 1982; Wendt et al., 1985] and Hannover University [e.g. Kanngieser, 1982; Torge and Kanngieser, 1980] in a point network and along an E-W profile more than 100 km long (Fig. 2). There are also many geodetic observations in the immediate region of the Krafla caldera taken by Icelandic geoscientists at shorter time intervals [Björnsson et al., 1985], dominantly relating to the inflation-deflation activity of the magma chamber, the effects of which fall off rapidly with distance from the caldera center. We did not treat this data set extensively. The more regional data used here are effected mainly by the widening of the fissure swarm and by deep magma flow. The effects of the crustal magma chamber are taken into account only in a qualitative manner in this context.

Unfortunately the point network of horizontal and vertical displacements and the gravity profile do not coincide. It is therefore not possible to make the model fully consistent. Furthermore the different data sets are not equally sensitive to the various questions.

The displacement network is used mainly to infer the three-dimensional source geometry of fissuring and dyke formation during the observational intervals. Supplementary data are direct observations of surface fissures. In addition the local seismicity during the rifting events is important for locating the new dykes.

The profile data on gravity and elevation across Northeast Iceland are used mainly to infer two-dimensional models of deep volume and density changes between 1975 and 1980. Short-wavelength changes hint to fissure swarms other than Krafla to have been effected

by density and/or volume changes (in the neighbourhood of the profile), too, during the episode, but it is mainly the long-wavelength data that are used here.

A further data set on elevation changes has been observed by Tryggvason [1986, 1987] who observed the tilt of Lake Mývatn southwest of the caldera in an intermediate distance range. The interpretation is relevant to how the deep magma flow and the shallow activity of the magma chamber and the fissures are related to each other. We take this information into account in a qualitative manner. Similarly we consider seismicity along the active zone [partly unpublished data due to P. Einarsson, pers. comm., 1986] and discuss its relevance for long-distance connections between Krafla and the plume center. Finally we take a look at Iceland as a whole as a hot spot and use the gravity anomaly to infer some aspects of the internal structure of the convecting plume.

3. Dyke Injection and Horizontal Displacements

We now take a closer look on the horizontal displacement data and the model of dyke injection in the fissure swarm. We shall first briefly review the theory.

Maruyama [1964] showed that opening a fissure generates a far-field deformation equivalent to that generated by a torque-free force couple acting in opposite directions [Mindlin and Cheng, 1950]. The rock body is approximated by an elastic half space. Such a model was applied by Pollard and Holzhausen [1979], Marquart [1983], and

Marquart and Jacoby [1985] with tensional stress acting on fissure walls.

The displacements in an elastic half-space with a free surface at z=0 as a result of the above force couple are [Mindlin and Cheng, 1950]:

$$u_x = -\frac{I_p x}{2\mu}\left\{\frac{1}{R_1^3} - \frac{3y^2}{R_1^5} + \frac{3-4\nu}{R_2^3} - \frac{3(3-4\nu)y^2}{R_2^5}\right.$$
$$-\frac{6cz}{R_2^5} + \frac{30czy^2}{R_2^7} - \frac{4(1-\nu)(1-2\nu)}{R_2(R_2+z_2)^2}$$
$$\left.+\frac{4(1-\nu)(1-2\nu)y^2}{R_2^3(R_2+z_2)^2} + \frac{8(1-\nu)(1-2\nu)y^2}{R_2^2(R_2+z_2)^3}\right\}$$

$$u_y = -\frac{I_p y}{2\mu}\left\{\frac{1-4\nu}{R_1^3} - \frac{3y^2}{R_1^5} + \frac{5-8\nu}{R_2^3} - \frac{3(3-4\nu)y^2}{R_2^5}\right.$$
$$-\frac{18cz}{R_2^5} + \frac{30czy^2}{R_2^7} - \frac{12(1-\nu)(1-2\nu)}{R_2(R_2+Z_2)^2}$$
$$\left.+\frac{4(1-\nu)(1-2\nu)y^2}{R_2^3(R_2+Z_2)^2} + \frac{8(1-\nu)(1-2\nu)y^2)}{R_2^2(R_2+Z_2)^3}\right\} \qquad (1)$$

$$u_z = -\frac{I_p z}{2\mu}\left\{\frac{z_1}{R_1^3} - \frac{3y^2 z_1}{R_1^5} + \frac{(3-4\nu)z_1}{R_2^3} - \frac{3(3-4\nu)y^2 z_1}{R_2^5}\right.$$
$$-\frac{6czz_2}{R_2^5} + \frac{30czy^2 z_2}{R_2^7} + \frac{4(1-\nu)(1-2\nu)}{R_2(R_2+Z_2)}$$
$$\left.-\frac{4(1-\nu)(1-2\nu)y^2}{R_2^3(R_2+Z_2)} - \frac{4(1-\nu)(1-2\nu)y^2)}{R_2^2(R_2+Z_2)^2}\right\}$$

with the point source intensity I_p, rigidity μ, Poisson's ratio ν, $R_1^2 = x^2+y^2+z_1^2$, $R_2^2 = x^2+y^2+z^2$, $z_1 = z-c$ and $z_2 = z+c$. The source acts upon the point $(0,0,c)$. The displacements caused by a source acting at $(x_c,0,c)$ are obtained by replacing in eq. (1) x by $x' = x - x_c$. To obtain the effects of an extended dyke-shaped source, we integrate the above solution with the coordinates (x ,0,c) across a vertical planar rectangle whose corners are given by the points $(x_o,0,z_o)$, $(x_1,0,z_o)$, $(x_1,0,z_1)$ and $(x_o,0,z_1)$. The formulae are complicated and given by Zdarsky[1987]. This procedure is not in all cases correct, because the point source solution involves the one upper free surface only and may degenerate if another shear-stress free surface of the fissure is added; however, in the present case the forces are normal to the fissure and the problem vanishes for reasons of symmetry. This has also been tested a posteriori [Zdarsky, 1987].

As a result a constant jump Δu_y across the source area is obtained which is related to its strength I_a by

$$\Delta u_y = 4\pi(1-2\nu)\frac{I_a}{\mu} \qquad (2)$$

Rigidity μ has apparently no influence on the shape of the displacement field, only on its magnitude, but Poisson's ratio ν influences both; we set $\nu = 0.25$.

Edge effects at the rim of the source in the form of stress peaks diminish rapidly with distance and are thus not critical to our results. The assumption of a homogenious half-space was also shown to be uncritical [Jacoby et al.,1983]. Fig. 3 shows the horizontal displacements of surface points caused by a two-dimensional fissure computed (with a Finite-Element-procedure [Zienkiewicz, 1977]) for varying Poisson's ratio of the half-space. As long as Poisson's ratio is less than 0.4 the displacements are very similar, only greater values lead to obvious differences. Seismic studies suggest that Pois-

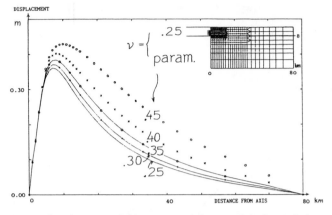

Fig. 3. Displacement fields computed for a model of an infinite dyke embedded in a layered elastic half-space. The parameters at the curves are Poisson's ratios varied in the lower layer.

son's ratio should be between 0.25 and 0.35 [Gebrande et al., 1980; Flóvenz, 1980]; thus the half-space model does not seem to introduce large errors.

The models were developed by taking rectangular sources of varying length, depth and location, then computing their individual displacement fields and finding their individual amplitudes or strengths by a least-squares fit to the observations. The non-linear effects of source position and geometry were not directly inverted but were determined by a manual trial and error procedure. The fit was considered satisfactory when the standard deviation was similar to the observational one.

Plots of enlarged displacement vectors observed in time intervals between the summer field seasons 1971, 1975, 1977, 1980 and 1982 are shown in Fig. 4. They are taken from Wendt et al. [1985] with additional data from K. Wendt [pers. comm., 1986]. For modelling the axis of the fissure swarm was held fixed as a reference, after the location of the axis had been optimized by a least-squares procedure on the assumption of symmetry of the displacement field about it [Zdarsky, 1987]. Fig. 5 presents the model results. The dykes that are limited vertically and horizontally are shown in map view (seen from ESE, prependicular to the swarm axis). Also sketched are the boundaries of the Krafla fissure swarm, of the Krafla caldera, and some of the earthquake swarms that occurred during the observational time intervals.

Most model dykes coincide with earthquake swarms as far as they are known to us to have occurred during respective deflation events. Since we had not taken into account the seismicity when modelling the dykes, our interpretation in these cases is strongly supported. One model dyke that lies north of the caldera where the observational network was dense enough for a reliable location, is related to the earthquake swarm of July 1978, and Fig. 6 shows its position in the relation to the epicenters and hypocenters [Einarsson and Brandsdóttir, 1980]; it lies fully within the swarm both horizontally and vertically. We note that another dyke is related to the deflation and eruption of July 1980 which flooded part of the fissure area and filled some of the fissures from above. Again, we had not taken this fact into account, but found that the depth extent of the model dyke had to be limited to the upper 20 m to achieve a good fit.

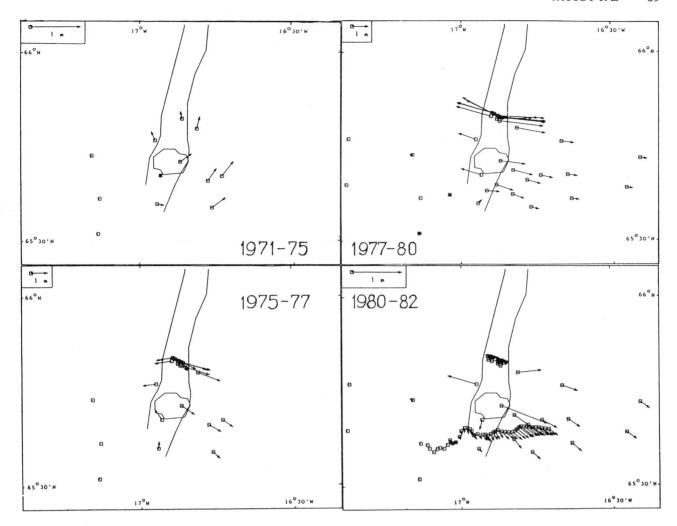

Fig. 4. The horizontal displacements observed in the network shown in Fig. 2 (or in parts of the network) for the time intervals between the summers of 1971, 1975, 1977, 1980, and 1982. Note the different scales of the displacement vectors.

Some dykes were found in the region of the magma chamber, and we believe that they poorly depict its relative inflation state. All of these model dykes except one indicate a volume increase. The exception falls into the time interval at the end of which the magma chamber was less inflated than initially (Fig. 7 and Björnsson [1985]). This is evidence that we cannot neglect the magma chamber in our modeling. Finally, a few model dykes outside the fissure swarm cannot be related to geology; they are probably an artefact of our modeling procedure.

We conclude that the horizontal displacement field contains considerable information to the development of rifting. The model dykes can often be located with an error of a few hundred meters and add constraints to the paths of magma motion to those provided by seismicity. The models should, however, be refined in the future to take account, for instance, of the magma chamber as a more compact (i.e. point) source. Also the "far field" more than 10 km from the fissure swarm must be modelled better.

4. Deep Flow - Gravity and Elevation Changes

The gravity and elevation changes, observed along the E-W profile (Fig. 2) are believed to express changes of mass or density and volume at depth. From the spatial extent of the time-varying anomaly we conclude that we see effects from the crust-mantle transition zone which in this part of Iceland is the lithosphere-asthenosphere transition at the boundary of the diverging plates.

The observations are presented in Fig. 8. The reference is at Akureyri more than 50 km west of Krafla. No change is assumed to have occurred there. The field surveys were in the summers of 1975 and 1980. The onset of rifting was in December 1975. Elevation and gravity changes prior to 1975 (surveys 1964/5 and 1970) were small, with little obvious relation to the geology, and were possibly not significant [Torge and Kanngieser, 1980; Altmann, 1987]. The horizontal displacements [Möller and Ritter, 1980; Wendt et al., 1985] between 1970 and 1975 did, however, show already a divergent

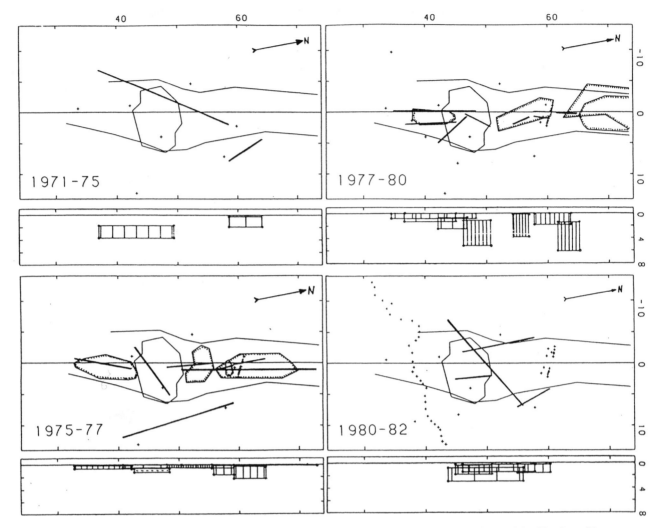

Fig. 5. Modeling results; for each time interval the best-fitting dykes are shown in mop view and in side view. The hatched regions are the sources of expansion, and the line density denotes the source strength.

component, perhaps related to the initial inflation of the magma chamber and the feeding flow.

The observational profile is long, extending more than 50 km to either side of the rift axis and is perpendicular to the axis. the profile passes about 10 km south of the center of activity in the caldera. Although it is not straight it gives little information on three-dimensional effects along the neovolcanic zone and its application to the whole region must therefore be regarded with caution.

At first sight (Fig. 8) we notice the graben subsidence and shoulder uplift of the Krafla fissure swarm and the mirror image gravity changes. The uplift is observed to about 30 km to either side, but with some asymmetry. A closer look reveals more differences. The most obvious difference is the broad gravity decrease near the center of the neovolcanic zone east of Krafla without much elevation change. There are also many short-wavelength (∼10 km) gravity changes which are each observed at a sufficient number of points so that they are taken to be real.

Application of the Free Air reduction leads to the temporal change of the Free Air anomaly

$$\Delta g_{FA} = \Delta g_{obs} + 0.3087 s^{-2} \Delta h \qquad (3)$$

Similarly, we obtain the temporal change of the Bouguer anomaly

$$\Delta g_{BA} = \Delta g_{FA} - 2\pi G \rho \Delta h \qquad (4)$$

(G = gravitational constant). For the density we have chosen $\rho = 2.6 g/cm^3$, appropriate for tholeiitic magma [Grönvold and Mäkinää, 1978]. Δg_{FA} and Δg_{BA} are included in Fig. 8. Evidently the Free Air reduction reverses the sign of the gravity effect: uplift (subsidence) is accompanied by increase (decrease) of the Free Air anomaly in the vicinity of the fissure swarm. No clear effects of the rifting are however visible in the change of the Bouguer anomaly; only on the eastern shoulder is Δg_{BA} negative (Bouguer reduction

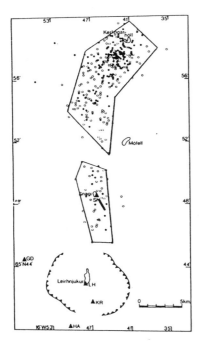

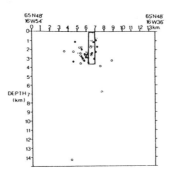

Fig. 6. Earthquake foci of the swarm accompanying the deflation event of July 1978 (from Einarsson and Brandsdóttir [1980]). One of our dyke models and the outlines of the seismically active regions are added; the other dykes we assumed are not plotted here.

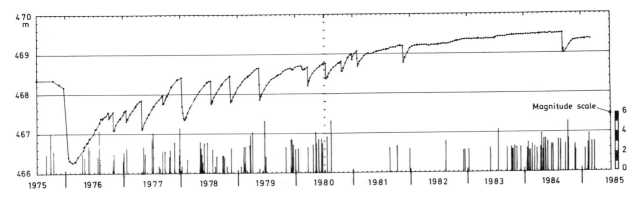

Fig. 7. The time variation of inflation/deflation of the Krafla magma chamber represented by the elevation of a benchmark (from Björnsson et al. [1985]). The observational periods are indicated by arrows. Seismicity in the Bárdarbunga region is added; vertical lines of different length indicate earthquakes of different magnitude (P. Einarsson, pers. comm., and data from Skjálftabréf, Science Institute, University of Iceland, 1977-1984. This is an unedited preliminary bulletin, and the data should thus be regarded with caution).

with $\rho = 2.6 g/cm^3$ "too strong"). The long-wavelength effects which coincide with the neovolcanic zone and the short-wavelength effects are not much changed by the reductions since the elevation changes are small.

The Bouguer anomaly change has a local extreme of -70 μgal, but there is a weak, though distinct trend with symmetry about the center of the neovolcanic zone with maxima at its margins and a broad flat low in between, perhaps with a small maximum in the middle (Fig. 8). The spectral density of Δg_{BA}, too, indicates that there is significant energy at wavelengths greater than 50 km (Fig. 9). This becomes clear in the low-pass filtered Δg_{BA} (Fig.10); the

filter has a cosine flank with its 3dB point at $\lambda = 98$ km. The observational uncertainty of the filtered (or averaged) anomaly is about 2 μgal instead of 10 μgal of the original observations. The filtered anomaly has a double amplitude of about 17 μgal. This is the effect we wish to interpret; it is small, but known with 10% accuracy, we believe.

Superimposed on the symmetrical "signal" there is a continuous decrease towards the east (or asymmetry). Although it is probably not an observational error, we shall not interpret it; it may be related to the asymmetry of location of the Krafla fissure swarm within the neovolcanic zone west of its axis. To fit a simple model, we

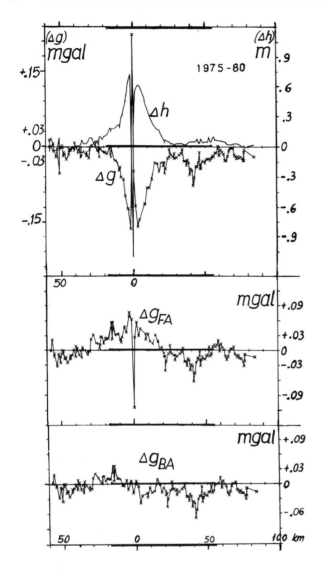

Fig. 8. The 1975 to 1980 change of the observed values of gravity Δg and elevation Δh and the change of the Free Air anomaly Δg_{FA} and the Bouguer anomaly Δg_{BA} along the profile from Akureyri to Vopnafjördur (see Fig. 2; from Kanngieser [1982]).

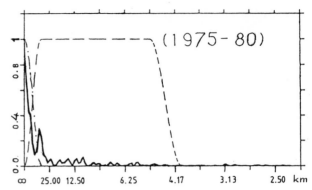

Fig. 9. The normalized power spectrum DGBA of Δg_{BA} versus wavelength; filter windows applied for isolating the short and long wavelength components (Fig. 10).

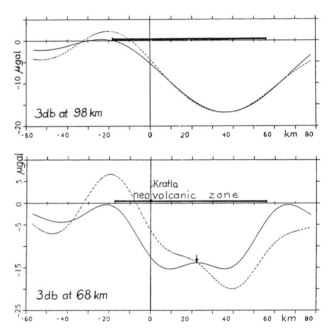

Fig. 10. Low-pass filtered and symmetricized change of the Bouguer anomaly; symmetry by averaging about the center: top and bottom: 3db point of low-pass filter at 98 and 68 km, respectively.

"symmetrize" the "signal" by averaging points of equal distances from the center of the neovolcanic zone. The symmetrical "signal" is shown in Fig. 10.

The interpretation is that hot magma moved into the axial region between the receding plates. We attempt to model this by an infinite horizontal line source of dilatation. For simplicity we assume that it is embedded in an elastic half-space. The model is a two-dimensional Mogi model [1958]. The dilatation generates a displacement field in the half-space, the divergence of which gives the negative density change for which we compute the surface gravity effects. Of course, the gravity effect of the surface deformation has to be taken into account in the Bouguer reduction so that we can compare the model

effect with the change in the Bouguer anomaly. While the density effect can be expressed analytically, the gravity effect is computed numerically. It turns out, however, that the model cannot simultaneously explain the surface deformation and the gravity effect, so that ultimately we have to revert to a fully numerical finite-element model with internal elastic structure by which both data sets can be explained satisfactorily.

The theory is briefly reviewed. The starting point is the density change:

$$-\frac{\delta\rho}{\rho} = \frac{\delta V}{V} = \underline{\nabla}\,\underline{u} = \frac{\delta u_y}{\delta y} + \frac{\delta u_z}{\delta z} \qquad (5)$$

The displacement field (u_y, u_z) id governed by the equilibrium at any point $(y, z) = x_i$

$$(\lambda + \mu)\frac{\delta\beta}{\delta X_i} + \mu\nabla^2 u_i + \rho X_i = 0 \qquad (6)$$

(with $\nabla^2 = \frac{\delta^2}{\delta y^2} + \frac{\delta^2}{\delta z^2}$; λ, μ = Lame's constants; $\beta = \underline{\nabla} \cdot \underline{u}$ and X_i = body force, i.e. gravity). Introduction of the Galerkin vector $\underline{f}$ leads to

$$2\mu\underline{u} = 2(1 - \nu)\nabla^2\underline{f} - \underline{\nabla}(\underline{\nabla} \cdot \underline{f}) \qquad (7)$$

(with = Poisson's ratio). After some algebra we obtain

$$u_y = \frac{I}{\mu}\left\{ \frac{y}{y^2 + (z - z_o)^2} + \frac{(3 - 4\mu)y}{y^2 + (z - z_o)^2} - \frac{4z(z + z_o)y}{(y^2 + (z + z_o)^2)^2} \right\} \qquad (8)$$

$$u_z = \frac{I}{\mu}\left\{ \frac{z - z_o}{y^2 + (z - z_o)^2} + \frac{4\nu - 1)z + (4\nu - 3)z_o}{y^2 + (z + z_o)^2} - \frac{4z(z + z_o)^2}{(y^2 + (z + z_o)^2)^2} \right\}$$

(with I = intensity of the line source and z_o = its depth). Differentiation according to eq. (5) gives

$$\delta\rho = \frac{4I\rho(2\nu - 1)}{\mu} \cdot \frac{(z + z_o)^2 - y^2}{(y^2 + (z * z_o)^2)^2} \qquad (9)$$

and the gravity effect or the change as the result of introduction of the line source is at the point (y_p, z_p) according to Fig.11

$$\Delta g_p = 2G \int_{y=-\infty}^{\infty} \int_{z=0}^{\infty} \frac{\delta\rho(y,z)(z - z_p)}{(y - y_p)^2 + (z - z_p)^2} dy dz \qquad (10)$$

The intensity I has a geological meaning if the source is a cylindrical magma channel of radius r_c under excess pressure Δp

$$I = \frac{\Delta p \cdot r_c^2}{2} \qquad (11)$$

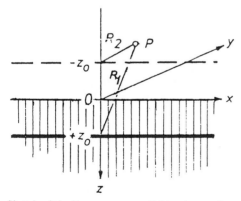

Fig. 11. Sketch of the line source parallel to the x-axis at depth z_o ; mirror image at $-z_o$.

The integral (10) is evaluated numerically. To circumvent computational instabilities we assume the observational level to be somewhat elevated above zero ground (1 km). The results are then compared with the observations after corresponding upward continuation. The physical parameters are chosen as $\rho = 2.6 g/cm^3$, $\nu = 0.26$, and $\mu = 3 \cdot 10^{10} Pa$.

The aim of modeling is to find the best fitting source strength and depth. Source strength is linearly related to gravity change and is computed by a least squares fit; depth, not linearly related, is systematically varied (between 5 and 35 km) and the minimum standard deviation is taken to give the most acceptable depth z_o . The results are intensity $I \approx 8 \cdot 10^7 Pa \cdot km^2$ and depth between 23 and 27 km ($z_o = 25$ km). The standard deviation of the fit is about 3 μgal or 20 % of the filtered observed effect. Fig. 12 compares the theoretical effect with the observed one. The intensity I can be translated via eq. (11) into an excess pressure at 25 km depth if a plausible radius r_c is assumed; for example if $r_c = 10$ km $\Delta p \approx 1.5 \cdot 10^6 Pa$ or about 0.2 % of the hydrostatic pressure at that depth.

If, with the above results, we compute the surface elevation change with eq. (8), z = 0, we find a maximum of +0.4 m instead of the observed change of up to 0.1 m (Fig. 12). A least-squares fit of the observed elevation changes directly gives an intensity $I \approx 10^7 Pa \cdot km^2$, much less than the above value. Variation of the physical parameters describing the half-space (elasticity, density) within plausible limits does not remove the discrepancy.

One reason for this discrepancy may be that the assumption of a homogeneous half-space is an oversimplification. Partial melt in the crust-mantle transition or in the elevated asthenosphere [Beblo and Björnsson, 1980] would soften this layer and enhance the deformation from the dilatation within it. This, in turn, would increase the gravity effect without necessarily amplifying the surface deformation of the strong crustal layer. The depth range of the soft layer (about 13 - 20 km) and our result ($z_o = 25$ km) suggest that the dilatation source is deep within the soft layer. Modification of our model is thus suggested. Visco-elasticity of the volume containing the source would, with time, have a similar effect as elastic "softness" has [Bonafede et al., 1986]. We shall model the soft layer approximately by assuming a large value for Poisson's ratio ($\nu = 0.38$) there.

An additional aspect is the positive gravity change near the margins of the neovolcanic zone, while the half-space model with a dilatational line source generates only very minute, but broad positive gravity flanks. It is to be tested whether the lateral limitation of the hot and soft zone explains the distinct gravity increase at the margins.

An analytical model with this complexity would be cumbersome at least if not intractable. We therefore revert to numerical finite element (FE) models. Fig. 13 shows the FE network we use and

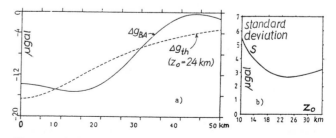

Fig. 12. a) Comparison of the best-fitting theoretical gravity change Δg_{th} with the observed — filtered (3db at 68 km) and symmetricized — change of the Bouguer anomaly Δg_{BA}. b) Variation of standard deviation s, as source depth z_o is varied.

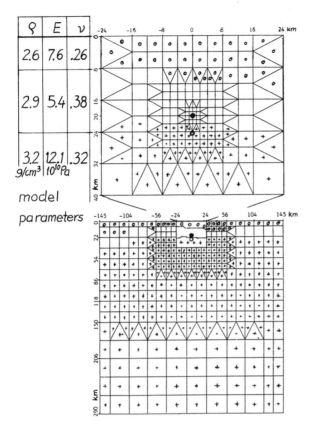

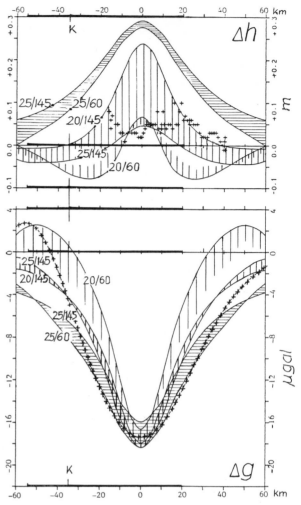

Fig. 13. Finite Element model network and the physical parameters assumed; the central part of the model with the source of expansion is enlarged.

Fig. 14. The Finite Element model results: elevation and gravity changes computed for 5 models and the observations (crosses, not shown near the Krafla fissure swarm for elevation); Δg not symmetricized, low-pass filtered with 3db point at 98 km. Horizontally hatched: homogeneous models; vertically hatched: layered model; parameters: source depth/model half width in km.

the spatial variation of μ, ν, and ρ. The model consists of three layers of variable depth and thickness [Björnsson, 1985]. Because of computer limitations our FE model is coarser and smaller in extent than that used in numerical integration of the gravity effect (eq. 10). Thus the FE results are less accurate and more qualitative. We compare several models. The upper boundary is always free. The sides are fixed at different distances from the source. The bottom is at always the same depth. For comparison with the analytical model a homogeneous FE model is included. Near the source the knot displacements are assumed as computed analytically from eq. 8 to make the models comparable to each other.

Fig. 14 presents the elevation and gravity changes computed and observed (except for the region strongly affected by the Krafla fissure swarm). The lateral restriction of the FE box suppresses the surface elevation somewhat, but not sufficiently. The three layered models show a much smaller surface uplift, even flanking subsidence. while in all models the gravity change is quite similar and in satisfactory agreement with the filtered observations. Although none of the models explains the observations perfectly, those where the model boxes have 145 (instead of only 60) km halfwidth and a source depth z_0 between 20 and 25 km (i.e. deep within the soft layer) render elevation changes that bracket the observations satisfactorily in view of the smallness and uncertainty of the effects.

In summary, a line source of dilatation along the axis of the neovolcanic zone, deep in the partially molten transition layer can

explain the observations along the long profile. The fit does, however, require elastic or visco-elastic layering. But we must not forget that we have no information on variations along the neovolcanic zone or on the validity of the two-dimensionality of the inferred structure.

There is a considerable easterly offset of the line source or magma flow at depth from the active crustal fissure swarm which may also explain the asymmetry of the gravity change Δg_{BA}. If our interpretation is correct the deep magmatic axis and plate boundary follows the geometric axis of the neovolcanic zone rather than the "magmatic axis" strung out by the central volcanoes. The Krafla fissure swarm, in this picture, has rifted not because it exactly follows the axis, but because it is the weakest part in a rather wide axial zone. If the deep dilatation was caused by the intrusion and flow of hot magma, excess pressure of the order of 10^6 Pa are inferred, quite compatible with the present observations and also with estimates of excess pressures in the Krafla magma chamber before eruption and rifting [Tryggvason, 1981].

5. Speculations on the Sub-Icelandic Plumbing System

Ideas, data, and interpretations presented in the previous sections are largely independent of each other. In this section we elucidate our introductory speculations on the provenance of the melt and on the connections between the melt at depth and the shallow dykes. Only one segment of the connection is fairly well established: the crustal magma chamber below Krafla caldera. But even its role is debated. Does all the melt ascend through the chamber or are there additional paths? How does the melt rise from the crust-mantle transition?

One indication is the chemical difference of the lavas inside and outside the caldera with an abrupt change at the marginal fault [K. Grönvold, pers. comm., 1984]. This suggests that the fault channels the flow or divides it into an upper, more differentiated and cooler stream surfacing inside the caldera and a lower, less differentiated and hotter one outside. Whether the deeper stream comes from a distinct and remote source or from deeper parts of the same one is not known, but there are some hints from the time behaviour of the 1984 eruption.

Tilt observations during and after this eruption out to 20 km from the center [Tryggvason, 1986] were made with 4 recording tiltmeters (3 within the caldera and 1 outside, 10 km to the south), with 18 optical levels (within 14 km from the center, June to December 1984, and with 6 lake level stations at Mývatn (between 11 and 20 km from the center during the same period). The tilt versus distance (from the inflation/deflation source) values deviate from what is computed for a point source at 2.6 km depth in an elastic half-space in such a way that the tilt maximum is sharper peaked and the distant flanks are broader beyond about 6 km distance. This suggests that the source may have a shallower and a deeper part or a noticeable vertical extent (see also Marquart [1983]).

The space-time tilt behaviour is another indication of deep flow. The eruption began on Sep., 4 with a very rapid deflation of the crustal reservoir, followed by renewed inflation while the eruption continued and even increased. Deflation subsequently resumed till Sep., 81, when the eruption ceased rather abruptly and completely. Inflation restarted then immediately and it continued at a decaying rate. The time periods Sep., 18 - 26 and Sep., 26 to Dec. 1984 were characterized by exponential decay of the inflation, first rapid, then slower. After Dec. 1984 inflation continued at a much slower and somewhat irregular rate. However, one may interpret the data by assuming a gradual decrease of the decay rate, slower than simply exponential, for the inflation. The behaviour may indicate that the magma inflating the crustal reservoir came through two or more reservoirs or an elongated one and that the flow was slowed by increased resistance. The most simple model is that of a dipole (inflation of the shallow reservoir and simultaneous deflation of a deeper one); Tryggvason [1986] demonstrated that indeed the space-time behaviour or tilt does support the model (inward tilting beyond 10 km distance for some time after Sep., 18 and simultaneous outward tilting of the close-in stations; only after Sep., 25, or early October all stations tilted outward). It seems clear that the magma inflating the crustal reservoir was drawn from increasingly greater depths or distances, but the geometry is not well constrained. Tryggvason's [1986] estimates were 8 km depth for the deflating reservoir during

the first week (Sep., 18 to 26) and 20 km afterwards till December. This picture fits fairly well with our own results, particularly with our deep dilatation source (for 1975 to 1980).

The 8 km reservoir down below Krafla is probably within the crust-mantle transition (based on magneto-telluric [Beblo and Björnsson, 1980; Arnason, 1981; Beblo et al., 1983] and reflection-seismic data [Zverev et al., 1980], see also: Björnsson [1980]; Marquart and Jacoby [1985]) and may naturally result from chemical differentiation in the vertical column. The silica-enriched crustal reservoir forms its bottom by solidified cumulates. At some depth below, however, temperature again exceeds the solidus of the transitional material so that melt forms another pocket which also provides the pathway from greater depth. "New" melt has intruded since 1975 and increased the pressure in the reservoirs. If new melt reached the crustal reservoir it probably settled below the differentiated lighter one, resulting in density layering. When the Krafla magma chamber burst the fissures were injected according to this layering, with the shallower injections escaping to the surface in the caldera and the deeper injections outside. This is also in agreement with the latest eruption. The main flow occurred outside the caldera (to the north), and when it already stopped, the pressure in the crustal (upper!) reservoir immediately began to rise (inflation) while deflation of the 8 km deep reservoir continued. This strongly suggests that the eruptions outside the caldera have been fed from the crustal reservoir, probably from its deeper parts.

As mentioned, inflation has continued at a very slow rate (as before 1984), while it should stop when pressure (i.e. the potential pressure or ambient minus hydrostatic) has equilibrated between the different reservoirs. This suggests that there is another - deep or more remote - source of melt. Tryggvason [1986] hints to the immense volume of partially molten anomalous upper mantle below Iceland, but should this not cause a steady inflow (and inflation) throughout time instead of the observed episodicity? There may be more than one answer. One could be that the transition zone is irregularly charged with melt blobs from the underlying anomalous mantle. Perhaps such a blob entered the zone below Bárdarbunga (deflation of a deeper "blob reservoir" with the accompanying thrust-type seismicity in the roof), as described in the introduction. The correlation between the Bárdarbunga seismicity and Krafla activity [Einarsson, 1987] could be explained this way. In this case Bárdarbunga would be the cause and Krafla the effect. Another possibility is that visco-elastic relaxation of the lithosphere-asthenosphere system after the initial crustal (i.e. lithospheric) rifting in the Krafla fissure swarm slowly permits more melt to rise between the receding "plates". The cause and effect relationship would then be reversed, i.e. as P. Einarsson suggested (see introduction), but the correlation between the Bárdarbunga seismicity and the Krafla activity would be equally well explained. As in any complex system, it is hardly possible to decide which is cause and which is effect. At present we are lacking the dicriminating data.

Finally we wish to relate the magmatic system to the Iceland plume. Why should blobs, e.g., rise particularly near its center? We think that it is because the anomalous upper mantle is in a state of internal small-scale convection with the hot, low-density, rising limb at the center, as sketched in the introduction. This is supported by the high volcanic production rates there and by the geochemistry of the Icelandic lavas. Another indication is the Bouguer-anomaly (BA) which has a rather narrow minimum (Fig. 15). In contrast, the gravity effect computed for the seismic model [RRISP Working

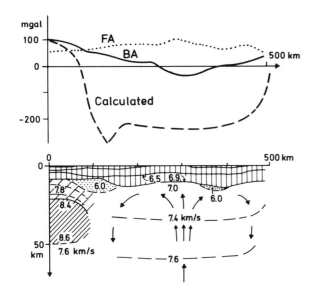

Fig. 15. A Bouguer anomaly profile across Iceland along RRISP profile I, compared with the gravity effect computed for the seismic model for the same profile shown below and with the assumption of a Birch-type velocity-density relationship [RRISP Working Group, 1980; Jacoby et al., 1983]. *BA* = Bouguer anomaly, *FA* = Free Air anomaly, vertical hatching: crust, oblique hatching: normal upper mantle of Reykjanes Ridge southeast flank, below Iceland: anomalous upper mantle.

Group, 1980; Gebrande et al., 1980; Jacoby et al., 1983] on the basis of the usual velocity-density systematics, has a very flat gravity minimum but a much greater amplitude. The discrepancy in shape of the anomaly is plausibly explained by a small melt content which decreases the velocities more than the densities. Partial melt will, however, coagulate particularly in the central upstream, and blobs are probably less "visible" to the seismic waves than to the gravity field. Thus, while gravity, of course, cannot prove this model, it is in very good agreement with it.

6. Concluding Remarks

We have presented a mosaic picture based on incomplete information which, while it is not a total overview, is useful for directing our further studies and has the following elements
-plume upwelling with a central convective upstream;
-progressive partial melting, blob coagulation, and upward percolation, preferentially near the center;
-flow divergence and pressure drop from the center and flow concentration along the "magmatic axis" marking the plate boundary;
-injection of blobs into the flow along the crust-mantle transition zone of melt accumulation (occurring repeatedly since about 1975);
-cascading pressure pulses from there along the magmatic axis (our line source) transmitted to shallower magma reservoirs at the intersections with fissure swarms (with volcanic complexes, calderas; altogether forming volcanic systems);
-pressure pulses may trigger the opening of a volcanic system, acting as a valve; only the "most ready" one will open at a time;
-pressurizing and inflation of reservoirs up to a critical point when

they break and magma is injected into fissures perpendicular to the minimum compression (deflation, rifting); chemically layered reservoirs will feed differing lavas;
-activity stops when a valve is sufficiently pressurized by the injections and the flow is blocked (Krafla may be at this stage now);
-visco-elastic relaxation after lithosphere rifting may prolong the rise of melt from depth.

It thus seems that we can now distinguish three intermediate reservoirs between the rising plume flow and the fissure injections and eruptions. It is remarkable that this picture is supported by many pieces of information. Tryggvason [1986] has condensed this picture into a simple flow model for the 1984 eruption which demonstrates the existence, as well as approximate depths and relative sizes of the intermediate reservoirs, in good agreement with the results presented here.

In an integral system as the one described it is difficult to call one part "active" and another "passive". Melt accumulation is needed to activate a rifting episode, as is a sufficiently slackened normal stress on the fissure walls. The melt will then rise where it finds the easiest route or where it can open a route. The surface vent need not be above the deep magma source.

The general situation of mid-ocean ridges may be similar. Melt can be produced in the ubiquitous asthenosphere by the pressure release caused by plate divergence where it moves over it. Temperature variations in the asthenosphere which may be caused by intruding plumes will result in different degrees of melting and volcanism with differing chemistry, leading to phenomena similar those in Iceland.

Acknowledgements. This paper was made possible by the generous supply of geodetic data from W. Torge and E. Kanngieser (Hannover), D. Möller, B. Ritter and K. Wendt (Braunschweig), as well as from several Icelandic colleagues. Our ideas grew in many helpful discussions, particularly with Páll Einarsson, Bryndís Brandsdóttir, Sveinbjörn Björnsson, Axel Björnsson, Gudmundur Pálmason, Karl Grönvold, Kristjan Saemundsson, Sigurdur Steinthorsson, Eysteinn Tryggvason and many others in Reykjavik, and Gaby Marquart (Uppsala). The constructively critical reviews by Páll Einarsson and George P. L. Walker greatly improved the paper, but of course we bear the responsibility for its faults. Computations were done at the computing centers of Frankfurt and Mainz universities. We greatfully acknowledge the help we received from many Icelanders, besides those already mentioned, particularly Stefán Sigurmundsson (Reykjavik) and Hjörtur Tryggvason (Húsavik) whose hospitality and guidance in the field strongly influenced our thinking about Iceland. Finally, this work would not have been possible without the financial support by Deutsche Forschungsgemeinschaft.

References

Altmann, U., Modellierung von Massenverschiebungen vor und während der Riftphase des Krafla-Spaltenschwarms in Nordisland anhand von Schwere- und Höhenänderungen, *Ph. D. thesis*, Universität Frankfurt, 1987.
Arnason, K., Magnetotellurische Messungen auf einem Profil über den Zentralvulkan Krafla in Nord-Ost-Island, *Diplomarbeit*, Universität München, 1981.
Beblo, M., and A. Björnsson, A model of electrical resistivity beneath NE Iceland, Correlation with temperature, *J. Geophys.*, *47*, 184-190, 1980.

Beblo, M., A. Björnsson, K. Arnason, B. Stein, and P. Wolfgram, Electrical conductivity beneath Iceland - Constraints imposed by magnetotelluric results on temperature, partial melt, crust and mantle structure, *J. Geophys.*, *53*, 16-23, 1983.

Björnsson, A., Dynamics of crustal rifting in NE-Iceland, *J. Geophys. Res.*, *90*, 10151-10162, 1985.

Björnsson, A., G. Björnsson, A. Gunnarson, and G. Thorbergsson, Breytingar á landhaed vid Kröflu 1974-1984 (in Icelandic), *Orkustofnun Report OS-85019/JHD-05*, Reykjavik, 1985.

Björnsson, A., K. Saemundsson, P. Einarsson, E. Tryggvason, and K. Grönvold, Current rifting episode in North Iceland, *Nature*, *266*, 318-323, 1977.

Björnsson, A., G. Johnsen, S. Sigurdsson, G. Thórbergsson, and E. Tryggvason, Rifting of the plate boundary in North Iceland 1975-1978, *J. Geophys. Res.*, *84*, 3029-3038, 1979.

Bonafede, M., M. Dragoni, and F. Quareni, Displacement and stress fields produced by a centre of dilatation and by a pressure source in a visco-elastic half-space: Application to a study of ground deformation and seismic activity at Campi Flegrei, Italy, *Geophys. J. R. Astron. Soc.*, *87*, 455-485, 1986.

Einarsson, P., Seismicity along the eastern margin of the North American plate, in press, 1987.

Einarsson, P., The anomalous mantle beneath Iceland and possible magma pressure connection between volcanoes (abstract), *Hawaii Symposium, How Volcanoes Work*, Hawaii, Jan. 1987.

Einarsson, P., and B. Brandsdóttir, Seismological evidence for lateral magma intrusion during the July 1978 deflation of the Krafla volcano in NE Iceland, *J. Geophys.*, *47*, 160-165, 1980.

Flóvenz, O. G., Seismic structure of the Icelandic crust above layer three and the relation between body wave velocity and the alteration of the basaltic crust, *J. Geophys.*, *47*, 211-220, 1980.

Gebrande, H., H. Miller, and P. Einarsson, Seismic structure of Iceland along RRISP profile I, *J. Geophys.*, *47*, 239-249, 1980.

Grönvold, K., Mäkipää, Chemical composition of the Krafla lavas 1975-1977, *Nordic Volcan. Inst., Univ. Iceland*, Reykjavik, 1978.

Jacoby, W., A. Björnsson, and D. Möller (Eds.), Iceland: Evolution, Active Tectonics, and Structure, *J. Geophys.*, *47*, 1-277, 1980.

Jacoby, W., U. Altmann, and G. Marquart, Structure and evolution of Iceland and the Reykjanes Ridge, *Rep. Abschlußkolloquium Meteor Expedition 45*, 67-83, Inst. Geophys., Univ. Hamburg, 1983.

Kanngieser, E., Untersuchungen zur Bestimmung tektonisch bedingter zeitlicher Schwere- und Höhenänderungen in Nordisland, *Wiss. Arb. Fachricht. Vermessungswesen, Dissertation, Nr. 114*, Univ. Hannover, 1982.

Marquart, G., Modellierung der Oberflächendeformationen im Gebiet des seismisch und tektonisch aktiven Krafla-Spaltenschwarms in Nordost-Island, *Berichte Inst. Meteorol. Geophys.*, Univ. Frankfurt, *51*, 1983.

Marquart, G., and W. R. Jacoby, On the mechanism of magma injection and plate divergence during the Krafla rifting episode in NE Iceland, *J. Geophys. Res.*, *90*, 10178-10192, 1985.

Maruyama, T., Statical elastic deformations in an infinite and semi-infinite medium, *Bull. Earthquake Res. Inst., Tokyo*, *42*, pp. 289-368, 1964.

Mindlin, R. D., and D. H. Cheng, Nuclei of strain in the semi-infinite solid, *J. Appl. Phys.*, *21*, 926-930, 1950.

Minster, J. B., and T. H. Jordan, Present day plate motions, *J. Geophys. Res.*, *83*, 5331-5354, 1978.

Mogi, K., Relations between the eruptions of various volcanous and the deformation of the ground surfaces around them, *Bull. Earthquake Res. Inst.*, *36*, 99-134, 1958.

Möller, D., and B. Ritter, Geodetic measurements and horizontal crustal movements in the rift zone in NE Iceland, *J. Geophys.*, *47*, 110-119, 1980.

Pálmason, G., and K. Saemundsson, Iceland in relation to the Mid-Atlantic Ridge, *Ann. Rev. Earth Planet. Sci.*, *2*, 26-50, 1974.

Pollard, D. D., and G. Holzhausen, On the mechanical interaction between a fluid-filled fracture and the earth's surface, *Tectonophysics*, *53*, 27-57, 1979.

Ritter, B., Untersuchungen geodätischer Netze in Island zur Analyse von Deformationen von 1965 - 1977, *Deutsche Geodät. Komm., Ser. C*, *271*, 1982.

RRISP Working Group, Reykjanes Ridge Iceland Seismic Experiment (RRISP 77), *J. Geophys.*, *47*, 228-238, 1980.

Schmeling, H., Partial melt below Iceland: A combined interpretation of seismic and conductivity data, *J. Geophys. Res.*, *90*, 10105-10116, 1985.

Steinthorsson, S., and W. R. Jacoby (Eds.), Crustal accretion in and around Iceland, *J. Geophys. Res.*, *90*, 9951-10192, 1985.

Torge, W. and E. Kanngieser, Gravity and height variations during the present rifting episode in northern Iceland, *J. Geophys.*, *47*, 125-131, 1980.

Tryggvason, E., Pressure variations and volume of the Krafla magma reservoir, *Nordic Volcanol. Inst.*, Rep. 8105, Reykjavik, 1981.

Tryggvason, E., Multiple magma reservoirs in a rift zone volcano: Ground deformation and transport during the September 1984 eruption of Krafla, Iceland, *J. Volcanol. Geotherm. Res.*, 1-44, 1986. Tryggvason, E., Mývatn lake level observations 1984-1986 and ground deformation during a Krafla eruption, *J. Volcanol. Geotherm. Res.*, *31*, 131-138, 1987.

Vogt, P. R., Plumes, sub-axial pipe flow, and topography along the mid-oceanic ridge, *Earth Planet. Sci. Lett.*, *29*, 309-325, 1976.

Walker, G. P. L., The structure of eastern Iceland, in *Geodynamics of Iceland and the North Atlantic area*, edited by L. Kristjansson, D. Reidel, pp. 177-188, Hingham, Mass., 1974.

Wendt, K., D. Möller, and b. Ritter, Geodetic measurements of surface deformations during the present rifting episode in NE-Iceland, *J. Geophys. Res.*, *90*, 10163-10172, 1985.

Zdarsky, H. Analytische Modelle für die Horizontalverschiebungen während der Riftepisode des Krafla-Spaltenschwarms in Nordost-Island, *Diplomarbeit*, Univ. Frankfurt, 1987.

Zienkiewicz, O. C., *The Finite Element Method*, 3rd expand. revised ed., McGraw-Hill, London, 1977.

Zverev, S. M., I. V. Litvinienko, G. Pálmason, G. A. Yaroshevskaya, N. N. Osokin, and M. A. Akhmetjev, A seismic study of the rift zone in northern Iceland, *J. Geophys.*, *47*, 191-201, 1980.